Counterfeit Integrated Circuits

Mark (Mohammad) Tehranipoor • Ujjwal Guin
Domenic Forte

Counterfeit Integrated Circuits

Detection and Avoidance

Springer

Mark (Mohammad) Tehranipoor
ECE Department
University of Connecticut
Storrs, CT, USA

Ujjwal Guin
ECE Department
University of Connecticut
Storrs, CT, USA

Domenic Forte
ECE Department
University of Connecticut
Storrs, CT, USA

ISBN 978-3-319-11823-9 ISBN 978-3-319-11824-6 (eBook)
DOI 10.1007/978-3-319-11824-6

Library of Congress Control Number: 2014960282

Springer Cham Heidelberg New York Dordrecht London

Printed on acid-free paper

Springer International Publishing AG Switzerland is part of Springer Science+Business Media (www.springer.com)

Dedicated to Maryam, Bahar, and Parsa
MT
Dedicated to my family
UG
Dedicated to my family
DF

Preface

Integrated circuits (ICs) and other electronic components form the foundation of the modern systems and infrastructures responsible for energy, finance, communication, defense, and much more. Over the last decade or so, increasing globalization has resulted in a dramatic increase in vulnerabilities within the electronic component supply chain. In particular, the rise of counterfeit ICs has become one of the most serious issues faced by industry, government, and society. Counterfeit ICs are literally a multibillion dollar business and growing at an unprecedented rate, impacting the profits of intellectual property (IP) holders as well as their corporate identities and reputation. Due to the widespread use of electronic components in our day-to-day lives—both directly and indirectly—counterfeit components also pose substantial threats to the health, safety, and security of the population at large.

This book is intended to serve as a resource for both beginners and experts in the counterfeit electronic components domain. For newcomers to the area, it shall introduce all of the necessary background material. This book aims to provide a comprehensive description of all different types of counterfeit ICs and the safety and security threat posed by these components. We believe a complete understanding of the detection of such components is a prerequisite if the community wants to stay ahead of the counterfeiters. The physical and electrical test methods described in this book provide guidance for the detection of these counterfeit components. We must also add design-for-anti-counterfeit (DFAC) measures into new ICs for a quick and easy counterfeit detection without the need to perform expensive physical and electrical test methods. This research-based book will provide the necessary road map for the government, industry, test labs, and academia throughout the world who are directly or indirectly impacted by this rampant attack of counterfeiting.

This book is organized into 12 chapters. The first chapter provides an introduction to counterfeit products in general. A significant portion of the total counterfeit trade belongs to counterfeit electronic products, which is supported by the data provided in this chapter. The next four chapters contain all the information one needs to know about counterfeit ICs. Chapter 2 presents a comprehensive taxonomy of counterfeit components, the vulnerabilities present in the different stages of the electronic component supply chain, and a brief overview of the current state of the art in the detection and avoidance of counterfeit electronic components. Chapter 3 illustrates all the defects and anomalies, namely, procedural, mechanical, environmental, and electrical defects, present in counterfeit components. Chapters 4 and 5 present all the physical and electrical tests currently available for the detection of these defects and thus counterfeit components. The challenges and limitations for existing tests and test procedures are also discussed in these chapters. These include the high test time and cost of detection, low confidence in detecting all counterfeit types, lack of automation, and so forth.

Starting in Chap. 6, we introduce recent work geared toward addressing many of the current issues. Chapter 6 focuses on improving the cost and effectiveness of existing tests. Specifically, it introduces the first test metrics to assess physical and electrical tests. A comprehensive framework is built upon these metrics to select the best set of test methods that maximizes counterfeit detection confidence under test time and cost constraints. Chapter 7 introduces two advanced physical inspection techniques to detect recycled and remarked ICs without involving subject matter experts in the decision-making process. Four-dimensional scanning electron microscopy and three-dimensional X-ray microscopy help to detect these counterfeit ICs in an effective and nondestructive way. Chapter 8 introduces several advanced electrical tests targeted specifically at two different types of recycled ICs—field programmable gate arrays (FPGAs) and application-specific integrated circuits (ASICs)—without performing the costly and time-consuming conventional physical and electrical tests.

Beginning in Chap. 9, we consider orthogonal approaches for addressing counterfeit detection and avoidance. Rather than relying on expensive test equipment and setups, these approaches integrate new test structures and primitives into the die and/or package (i.e., design-for-anti-counterfeit (DFAC)) to actively target different counterfeit types with much greater ease. First, Chap. 9 introduces several low-cost combating die and IC recycling (CDIR) structures to detect recycling in a wide range of electronic component types (from large digital ICs to small analog and discrete components). Chapter 10 discusses the IP theft problem and gives an overview of passive watermarking techniques capable of providing proof of IP authorship with high confidence. Chapter 11 discusses the counterfeit threats associated with untrusted foundries and assemblies along with countermeasures recently proposed, such as Connecticut Secure Split Test (CSSS), to prevent overproduced, cloned, and out-of-spec/defective ICs from being introduced into the supply chain. Finally, Chap. 12 introduces package IDs based on encrypted QR codes, DNA markings, nanorods (NR), and coating physical unclonable functions (PUFs), which can be potentially implemented in all the component types to detect recycled, remarked, overproduced, and cloned ICs.

Storrs, CT, USA
December, 2014

Mark (Mohammad) Tehranipoor
Ujjwal Guin
Domenic Forte

Acknowledgments

The authors would like to acknowledge the National Science Foundation (grants CCF-1423282 and CNS 1344271), Missile Defense Agency (MDA), Honeywell Inc., Comcast, and SAE G-19A group for supporting our projects in the area of counterfeit integrated circuit detection and prevention. We would also like to thank the following individuals for their valuable contributions and feedback to the book:

- **Bicky Shakya** for his contributions to Chaps. 2 and 10 and proofreading the book.
- **Dr. Navid Asadizanjani** for his contributions to Chaps. 3 and 7.
- **Dr. Xuehui Zhang** for her contributions to Chaps. 8 and 9.
- **Halit Dogan** for his contributions to Chap. 8.
- **Tauhidur Rahman** for his contributions to Chap. 11.
- **Dr. Sina Shahbazmohamadi** for his contributions to Chap. 7.
- **Daniel DiMase** of Honeywell Inc. for providing valuable feedback on Chaps. 2, 3, 4, and 6.
- **Steve Walters** of Honeywell Inc. for providing valuable feedback on Chaps. 3 and 6.
- **Mike Megrdichian** for providing valuable feedback on Chaps. 3 and 6.
- **Sultan Lilani** of Integra Technologies for providing valuable feedback on Chap. 5.
- **Emma Burris-Janssen** for proofreading the book.

Contents

Acronyms

AACF	Areal Autocorrelation Function
ABS	Anti-lock Braking System
ADNAS	Applied DNA Sciences
AES	Advanced Encryption Standard
AF	Antifuse
AF-CDIR	Antifuse-Based CDIR
ASIC	Application Specific Integrated Circuit
ATE	Automatic Test Equipment
BIS	Bureau of Industry and Security
BSE	Back-scattered Secondary Electron
BGA	Ball Grid Array
CAF	Clock AF
CAM	Content Addressable Memory
CBP	Customs and Border Protection
CCAP	Counterfeit Components Avoidance Program
CDC	Counterfeit Defect Coverage
CDIR	Combating Die and IC Recycling
CGA	Column Grid Array
CL	Confidence Level
CLB	Configurable Logic Block
CMOS	Complementary Metal Oxide Semiconductor
CoC	Certificates of Conformance
COTS	Commercial Off The Shelf
CPA	Counterfeit Prevention Authentication
CSB	Complete Scrambling Block
CSST	Connecticut Secure Split-Test
CT	Computed Tomography
CTC	Counterfeit Type Coverage
CTI	Components Technology Institute
CUA	Circuit Under Authentication
DA	Dynamic Assessment

DF	Defect Frequency
DFAC	Design for Anti Counterfeit
DFT	Design for Testability
DHS	Department of Homeland Security
DI	Decision Index
DIP	Dual In-line Package
DM	Defect Mapping
DNA	Deoxyribonucleic Acid
DOD	Department of Defense
DSP	Digital Signal Processor
ECID	Electronic Chip ID
EDS	Energy Dispersive Spectroscopy
EFR	Early Failure Rate
EMC	Epoxy Molding Compound
ESD	Electrostatic Discharge
EOS	Electrical Overstress
ERAI	Electronic Resellers Association International
ETD	Everhart-Thornley Detector
EVI	External Visual Inspection
FBI	Federal Bureau of Investigation
F-CDIR	Fuse-Based CDIR
FIB	Focused Ion Beam
FPGA	Field Programmable Gate Array
FPROM	Field Programmable Read Only Memory
FSM	Finite-State Machine
FTIR	Fourier Transform Infrared
FUT	FPGA Under Test
GDSII	Graphic Database System II
GIDEP	Government-Industry Data Exchange Program
HCI	Hot Carrier Injection
HD	Hamming Distance
HDL	Hardware Description Language
HIC	Humidity Indicator Cards
HM	Hardware Metering
IC	Integrate Circuit
ICC	International Chamber of Commerce
ICE	Immigration and Customs Enforcement
ID	Identification Number
IDEA	Independent Distributors of Electronics Association
IO	Input/Output
IP	Intellectual Property
IPA	Isopropyl Alcohol
IPR	Intellectual Property Rights
LFSR	Linear Feedback Shift Register
LUT	Look Up Table

MBB	Moisture Barrier Bag
MC	Monte Carlo
MCS	Monte Carlo Simulation
MISR	Multiple Input Signature Register
MPS	Ministry of Public Security
MSD	Moisture Sensitive Device
MSRP	Manufacturer's Suggested Retail Price
NBTI	Negative Bias Temperature Instability
NCD	Not Covered Defect
NR	Nanorods
NRE	Non Recurring Expense
OCM	Original Component Manufacturers
OECD	Organization for Economic Cooperation and Development
OPO	Original Primary Outputs
OTE	Office of Technology Evaluation
OTP	One Time Programmable
PBGA	Plastic Ball Grid Array
PCA	Principal Component Analysis
PCB	Printed Circuit Board
PIN	Part or Identifying Number
PLCC	Plastic Leaded Chip Carrier
PMU	Parametric Measurement Unit
PSB	Partial Scrambling Block
PUF	Physical Unclonable Functions
QR	Quick Response
R&D	Research and Development
RAM	Random Access Memory
RBF	Radial Basis Function
RE	Reverse Engineering
RFID	Radio Frequency Identification
RO	Ring Oscillator
ROHS	Restriction Of Hazardous Substances
RTL	Register Transfer Level
SA	Static Assessment
SAF	Signal Transition AF
SAM	Scanning Acoustic Microscopy
SB	Scrambling Block
SBCU	Scrambling Block's Controlling Unit
SEM	Scanning Electron Microscopy
SME	Subject Matter Expert
SiC	Silicon Carbide
SNR	Signal to Noise Ratio
SOA	Simple Outlier Analysis
SOC	System on Chip
SoW	Statement of Work

SRAM	Static Random Access Memory
SST	Secure Split Test
SVM	Support Vector Machine
TC	Target Confidence
TDDB	Time Dependent Dielectric Breakdown
TEM	Transmission Electron Microscope
TL	Tier Level
TRN	True Random Number
TRNG	True Random Number Generator
UCD	Under Covered Defect
UPC	Universal Product Code
VSIA	Virtual Socket Interface Alliance
WHO	World Health Organization
XRF	X-Ray Fluorescence
XRM	X-Ray Microscope

Chapter 1
Introduction

How confident are you that the Louis Vuitton handbag you bought on eBay is genuine? How can you be certain that the medication you're taking is free from harmful chemicals? How do you know that the components in your laptop were produced and inspected by a reliable manufacturer? With networks of production and consumption becoming increasingly globalized, the pressure to address problems associated with counterfeiting is incredibly intense.

While the scope of the counterfeit trade is difficult to assess due to its largely clandestine nature, there have been many recent—and troubling—incidents when the counterfeit trade and its victims have surfaced before the public eye. In 2006, the World Health Organization (WHO) estimated that between 10 % and 30 % of the medications circulating in developing countries were counterfeit [1, 2]. This can literally be a life-and-death problem for consumers. For instance, cough syrup containing ethylene glycol was identified as being responsible for the deaths of hundreds of people in Panama and the Dominican Republic in the span of less than a week [2]. From highly visible status symbols to dangerously invisible parts of the systems we rely on for our health and safety, counterfeiting is a far-reaching problem that requires a coordinated response.

In recent years, the "business" of counterfeiting has changed from the piecemeal production of inferior goods in small, clandestine workshops to the coordinated and sophisticated production of goods that are becoming harder and harder to differentiate from "the real thing." A case in point is that of Japanese electronics giant NEC: when NEC found out that pirated keyboards, CDs, and DVDs bearing the company logo were being circulated on the black market in Beijing and Hong Kong, they thought that they had a run-of-the-mill pirating threat on their hands [3]. What initially seemed like a routine intellectual property (IP) concern, however, turned out to be an ambitious counterfeiting campaign. Instead of just faking items sold by NEC, the counterfeiters were faking the entire company. The NEC counterfeiters established what was effectively a parallel brand where they not only copied NEC's product line but also developed their own range of consumer

M. Tehranipoor et al., *Counterfeit Integrated Circuits*,
DOI 10.1007/978-3-319-11824-6_1

electronic products. The counterfeiters ran this parallel business like any normal business: they had business cards printed, commissioned new product research and development, signed production and supply orders, and issued official-looking warranty and service documents. As president of International Risk (a Hong Kong-based company hired by NEC to investigate the piracy) Steve Vickers observed, the NEC case shows how drastically piracy is evolving. It has, in Vickers' words, "gone from often shoddy copying of brands to highly coordinated operations of production and marketing."

Though counterfeiting is not a problem specific to our historical moment, the stakes of counterfeiting are especially high in an age where one faulty link in the chain of systems that protect and enmesh us can damage the lives of countless people around the world. The high stakes of modern counterfeiting are particularly evident in the case of "Operation Network Raider", a domestic and global initiative that aims to control the illegal distribution of counterfeit network hardware [4]. As of 2010, this initiative, which is a collaborative effort between the Federal Bureau of Investigation (FBI), Immigration and Customs Enforcement (ICE), Customs and Border Protection (CBP) and international agencies such as China's Ministry of Public Security (MPS), has led to 30 felony convictions and the seizure of nearly $143 million worth of counterfeit network hardware. In a majority of the incidents, a bulk of the seized counterfeit hardware came in the form of Cisco networking equipment, which was intended for use in US military computer networks and IT infrastructures for firewall protection and secure communication. In a separate case of counterfeiting, the CBP and ICE also made more than 1,300 seizures of semiconductor devices that were falsely marked as military/aerospace grade and were also affixed with trademarks of reputed semiconductor companies.

1.1 History of Counterfeiting

Far from being unique to our own time, counterfeiting has grown in tandem with human civilization. Wherever there are marks that authenticate, there are bound to be those that imitate in order to command the authority of the original, authenticating mark. In Babylon and ancient Egypt, for instance, priests hoped to increase their own legitimacy and proceeds by placing inscriptions from earlier civilizations on their monuments, thereby creating a "false authority".

Both trademarks and the counterfeiting of those trademarks stretch back into ancient times, with Pliny the Elder describing the popularity of counterfeit coins as collector's items within Roman society. The use of counterfeit coinage for illicit trade such as smuggling and certain types of foreign trade was the norm in sixteenth- and seventeenth-century Genoa. One of the most famous examples of coinage counterfeiting took place in Renaissance France when supporters of the Pope directed parallel minting in order to undermine the authority of France's Protestant king [2].

While the counterfeiting of money is—even now—perhaps the best-known form of counterfeit activity, evidence suggests that product counterfeiting may have been an even older form of illicit trade. During the first three centuries of the Roman Empire, oil lamps were marked with the FORTIS brand name, which scholars speculate was widely copied, given the widespread use of this stamp on many artifacts.

In recent years, the scope and magnitude of counterfeiting has exploded in size and volume. In July 2007, the Public Security Bureau of China and the FBI found $2 billion worth of counterfeit Microsoft software, including 19 versions of products in 11 languages, in a warehouse in southern China where workers assembled disks, authenticated materials and manuals, and prepared them for shipping, in what was probably the biggest counterfeit software bust in history. These counterfeit products turned up in 36 countries across six continents [5].

Thus, from the days of the ancient Rome to the rampant counterfeiting of products in the twenty-first century, counterfeiting has constantly evolved along with the advances in manufacturing and trade and is indeed an issue that needs to be addressed.

1.2 Counterfeit Products

A wide variety of products we use in our day-to-day lives are subjected to counterfeiting. Figure 1.1 shows a broader category of counterfeit products, consisting of such things as luxury goods, pharmaceutical products, and electronics. Among

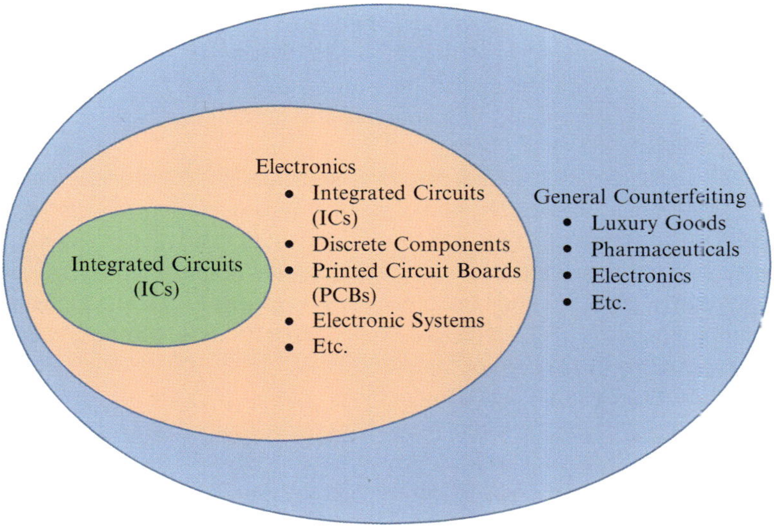

Fig. 1.1 Counterfeit products

them, electronics remain one of the major counterfeit products, where components such as integrated circuits (ICs), discrete components, and printed circuit boards (PCBs) are frequently counterfeited. Considering the severe impact that counterfeit ICs can have on the safety and security of hardware underlying today's information systems and keeping in mind the scope of this book, we will describe the detection and avoidance of these ICs in the successive chapters.

1.3 Counterfeits: A Trillion Dollar Market and Beyond

The growing incidence of counterfeit products constitutes a major challenge to the government, businesses, and consumers because it poses serious economic and safety concerns. A counterfeit product is "*any manufacturing of a product which so closely imitates the appearance of the product of another to mislead a consumer that it is the product of another*" [6]. Over the past few decades, these products have expanded rapidly across the globe due to globalization. Almost every country has become a part of the counterfeit trade. However, China, Korea, Chinese Taipei, Hong Kong (China), and the Philippines were the top five suppliers of counterfeit goods to the US in 1997, according to the data provided by US and EU States customs service [6]. With China as the single largest source economy, Asia remains the largest source for counterfeit and pirated products [7].

An accurate estimate of the size and value of the counterfeit market remains a mystery due to the clandestine nature of counterfeiting. The closest estimate of the size of the counterfeit market is made by an extrapolation from the amount of goods seized by the police and customs authorities. In a 2007 report, the Organization for Economic Cooperation and Development (OECD) asserted that "while the overall magnitude of counterfeiting and piracy cannot be easily measured, estimates of the role that counterfeit and pirated products are playing in international trade are possible" [7]. Based on their model, they estimated that up to US$200 billion of international trade was pirated and counterfeited in 2005, increasing to US$250 billion in 2007 [8].

In 2001, the International Chamber of Commerce (ICC) estimated that 5–7 % of world trade was in counterfeit goods and that the counterfeit market was worth $350 billion [2]. Based on 2008 data, it was estimated that counterfeit and pirated products could account for as much as US$650 billion per year globally. Due to the rapid increase in counterfeiting and piracy, it was predicted that the global market for counterfeit goods is likely to be more than double to US$1.7 trillion by 2015 [9]. Table 1.1 shows the breakdown of these estimates.

In the US, the Customs and Border Protection (CBP) agency is responsible for the seizure of counterfeit and pirated products that are imported and could possibly infringe US patents, trademarks, copyrights and other forms of intellectual property. Recent data from the CBP shows that the number of Intellectual Property Rights (IPR) seizures in 2013 increased by nearly 7 % from 2012. Table 1.2 shows the seizure of counterfeit goods by the U.S. Department of Homeland Security

Table 1.1 Estimate of the total value of counterfeit and pirated products [9]

OECD category	Estimate (2008 data)	Estimate (2015)
Internationally traded counterfeit and pirated products	$285–360 billion	$960 billion
Domestically produced and consumed counterfeit and pirated products	$140–215 billion	$570 billion
Digitally pirated products	$30–75 billion	$240 billion
Total	$455–650 billion	$1.77 trillion

(DHS). The People's Republic of China remains the primary source for producing counterfeit and pirated goods. It represents approximately 68 % ($1.1 billion) of all IPR seizures based on the manufacturer's suggested retail price (MSRP) in 2013.

Table 1.3 breaks down the seizure of counterfeit and pirated products by commodity type, revealing that, between FY 2012 and FY 2013, the counterfeit trade in computers and accessories went from $34,710,624 MSRP to $47,731,513 MSRP. Similar growth can also be seen in the counterfeit trade of consumer electronics and parts, which, as of FY 2013, was ranked behind only handbags, wallets, watches, and jewelry in its presence in the counterfeit market. More generally, total seizures increased 38 %, from $1,262,202,478 to $1,743,515,581 between FY 2012 and FY 2013.

In response to recent growth in the counterfeit trade, the first joint IPR enforcement operation between CBP and China Customs was coordinated and resulted in a staggering seizure of 1,735 shipments and the removal of more than 243,000 counterfeit consumer electronic products from the electronics supply chain. Also, in collaboration with French Customs, CBP completed Operation Core Systems, which resulted in the seizure of 480 shipments of potentially harmful counterfeit electronic components [10].

1.4 Counterfeit Electronics: An Emerging Threat

Counterfeit electronics pose a significant threat to the government and industrial sectors of the economy because they undermine the security and reliability of critical systems and networks. They have a negative impact on corporate identity and reputation, and they can trigger massive revenue losses. Due to the widespread use of electronic components in our day-to-day lives—both directly and indirectly—counterfeit components also pose major threats to the health, safety, and security of the population at large. For example, the failure of a pacemaker due to a counterfeit component can potentially take someone's life. Similarly, the anti-lock braking system (ABS), which is found in most cars today and is controlled by sensors and electronics, could possibly fail due to the use of counterfeit components. This not only causes reliability issues, it could potentially lead to life-threatening accidents.

Table 1.2 Seizure of counterfeit and pirated products by country [10]

FY 2013 source economy	Estimated MSRP ($)	Percent of total	FY 2012 source economy	Estimated ($)	Percent of total
China	1,180,919,064	68 %	China	906,206,684	72 %
Hong Kong	437,538,041	25 %	Hong Kong	156,337,345	12 %
India	20,683,669	1 %	Singapore	9,385,173	1 %
Korea	6,308,434	Less than 1 %	India	7,020,939	1 %
Singapore	5,065,398	Less than 1 %	Taiwan	4,500,610	Less than 1 %
Vietnam	4,406,367	Less than 1 %	Canada	4,236,359	Less than 1 %
Taiwan	3,975,422	Less than 1 %	France	4,221,443	Less than 1 %
Great Britian	2,421,034	Less than 1 %	Peru	2,760,392	Less than 1 %
Bangladesh	1,914,318	Less than 1 %	Mexico	2,673,976	Less than 1 %
Pakistan	1,335,728	Less than 1 %	Germany	2,280,520	Less than 1 %
All other economies	78,948,105	5 %	All other economies	162,579,037	13 %
Total FY 2013 est. MSRP	1,743,515,581		Total FY 2012 est. MSRP	1,262,202,478	
Number of seizures	24,361		Number of seizures	22,848	

Table 1.3 Seizure of counterfeit and pirated products by commodity type [10]

FY 2013 commodity	Estimated MSRP ($)	Percent of total	FY 2012 commodity	Estimated MSRP ($)	Percent of total
Handbags/wallets	700,177,456	40 %	Handbags/wallets	511,248,074	40 %
Watches/jewelry	502,836,275	29 %	Watches/jewelry	186,990,133	15 %
Consumer electronics/parts	145,866,526	8 %	Wearing apparel/accessories	133,008,182	11 %
Wearing apparel/accessories	116,150,041	7 %	Consumer electronics/parts	104,391,141	8 %
Pharmaceuticals/personal care	79,636,801	5 %	Footwear	103,365,939	8 %
Footwear	54,886,032	3 %	Pharmaceuticals/personal care	82,997,515	7 %
Computers/accessories	47,731,513	3 %	Optical media	38,404,732	3 %
Labels/tags	41,768,528	2 %	Computers/accessories	34,710,624	3 %
⋮	⋮	⋮	⋮	⋮	⋮
Total FY 2013 MSRP	1,743,515,581		Total FY 2012 MSRP	1,262,202,478	
Number of seizures	24,361		Number of seizures	22,848	

A pilot could lose control of an airplane, jeopardizing the lives of all on board. A rogue nation could even disable air defense systems with the help of counterfeit components.

In addition to the impact on public safety and security, counterfeit components could also cause significant damage to the economy. For example, semiconductor companies spend billions of dollars every year to develop technologies, manufacture products, and provide support for the products they create. In contrast, counterfeiters spend minimal money on developing technologies. Instead, counterfeiting practices allow private individuals to remake an existing product for their own benefit, which only hinders the research and development of new products. Also, as the counterfeiters do not take responsibility for their counterfeit components, the failure of these components damages the corporate reputation of the original component manufacturers (OCMs). In many cases, the OCM can even bear the financial responsibility and logistics of replacing the failed components.

It is also important to assess why these counterfeit incidents are on the rise. In the United States, only 25 % of electronic waste was properly recycled in 2009 [11]. This huge resource of e-waste allows counterfeiters to pile up an extremely large supply of counterfeit components. Counterfeiters recycle electronic components from this e-waste and sell them in the open market as if they were new or even of a superior grade (for example, commercial grade components are sold as military or space grade components). In addition to that, as the complexity of electronic systems and their components have grown significantly over the past few decades, these components have been increasingly assembled (fabricated) globally to reduce production costs. For example, large foundries located in different countries can offer lower prices to the design houses. Figure 1.2 reflects the trust and security

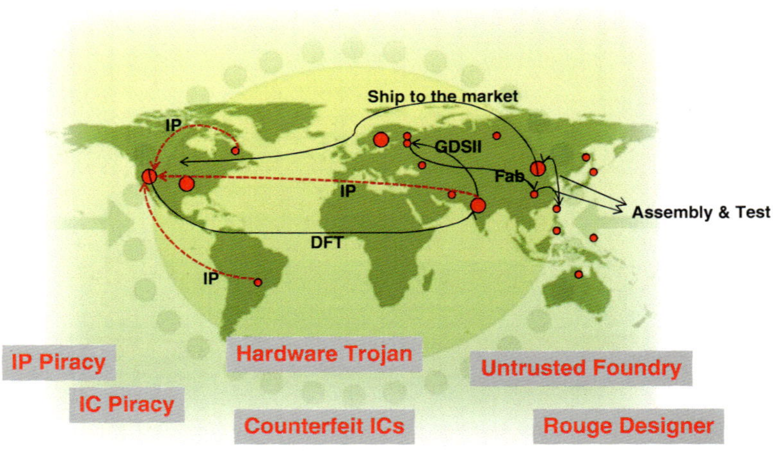

Fig. 1.2 Trust and security issues due to globalization

issues evident during the design and fabrication of electronic components. The designers use different IPs, collected from all across the globe, in their designs. It is extremely challenging and even impossible in some instances to validate their authenticity. The design-for-testability (DFT) for these integrated circuits is often inserted by third parties located in different places. Untrusted foundries and assemblies can also be capable of selling extra components outside of the number they were contracted to manufacture. Thus, this complex supply chain leads to an illicit market willing to undercut competition with counterfeit parts.

Due to the complex nature of the component supply chain, it is impossible to estimate the size of the actual semiconductor counterfeit market. Most estimates are based on data derived from the number of seized or detected components. Since a large portion of counterfeit semiconductor components could be circulating in the market, exact estimates are difficult to make. However, in the following section, we will present the "defense industrial base assessment for counterfeit electronics", a report prepared by the U.S. Department of Commerce that presents detailed statistics on counterfeiting to give us an idea of the magnitude of counterfeiting and to help us analyze how deeply this illegal activity is rooted in the supply chain.

1.4.1 Defense Industrial Base Assessment: Counterfeit Electronics

In June 2007, the Bureau of Industry and Security's (BIS) Office of Technology Evaluation (OTE) conducted a defense industrial base assessment of counterfeit electronics, as they suspected that counterfeit parts were infiltrating the Department of Defense's (DOD) supply chain and were, consequently, affecting the reliability of U.S. weapons systems. The primary objectives of this task were "*to assess: levels of suspected/confirmed counterfeit parts; types of devices being counterfeited; practices employed in the procurement and management of electronic parts; record keeping and reporting practices; techniques used to detect parts; and best practices employed to control the infiltration of counterfeits*" [12]. A total of 387 companies participated in this survey, which mainly focused on discrete electronic components, microcircuits, and circuit board products during a period stretching from 2005 to 2008.

It was found in the report that a majority of the OCMs found counterfeit versions of their components in the component supply chain. Table 1.4 shows the level to which the OCMs encountered counterfeit products. Around 46 % (18 out of 39) of the OCMs were the manufacturer of discrete components, whereas, 55 % (24 out of 44) of the OCMs were the producers of microcircuits (integrated circuits).

Table 1.4 Companies encountering counterfeit electronics [12]

Type of company	Encounter counterfeits	Did not encounter counterfeits	Total
Discrete electronic components	18	21	39
Microcircuits	24	20	44
Total	42	41	83

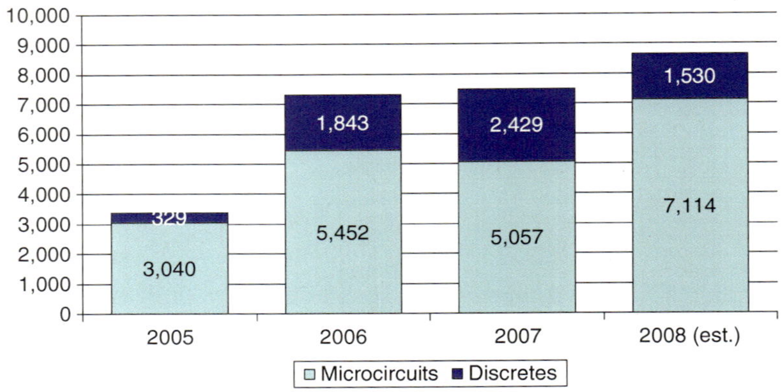

Fig. 1.3 Total counterfeit incidents—OCMs (2005–2008) [12]

This assessment also presented a steady increase in counterfeit incidents encountered by the OCMs. The number of counterfeit incidents grew by more than 150 % during the period from 2005 to 2008 (shown in Fig. 1.3). The number of incidents for discrete components soared to around 365 % during this same period, whereas they more than doubled for microcircuits.

Figure 2.1 shows the breakdown of counterfeit incidents for different parts that were reported in the assessment. Electromechanical components, thyristors, and capacitors are more vulnerable to counterfeiting, and they constituted around a quarter of the total counterfeit incidents for discrete parts (in Fig. 1.4a). Counterfeit microprocessors accounted for the most reported category of incidents within the microcircuit category (in Fig. 1.4b).

The report also touched on the resale value of counterfeit components. Counterfeiters do not always target high-end components with high resale values. Data from the report shown in Fig. 1.5 indicates that the resale value of a counterfeit part could be as low as a few pennies. The parts most often reported as counterfeit had resale values ranging from $0.11 to $500. Only a few high-end, costly parts with resale values of thousands of dollars were reported.

The report [12] also indicated that the majority of OCMs encountered counterfeit discrete components and microcircuits that were significantly different from their original counterparts. Most of the discrete components were fake and ultimately failed to produce correct responses. The rest mostly belonged to the "working

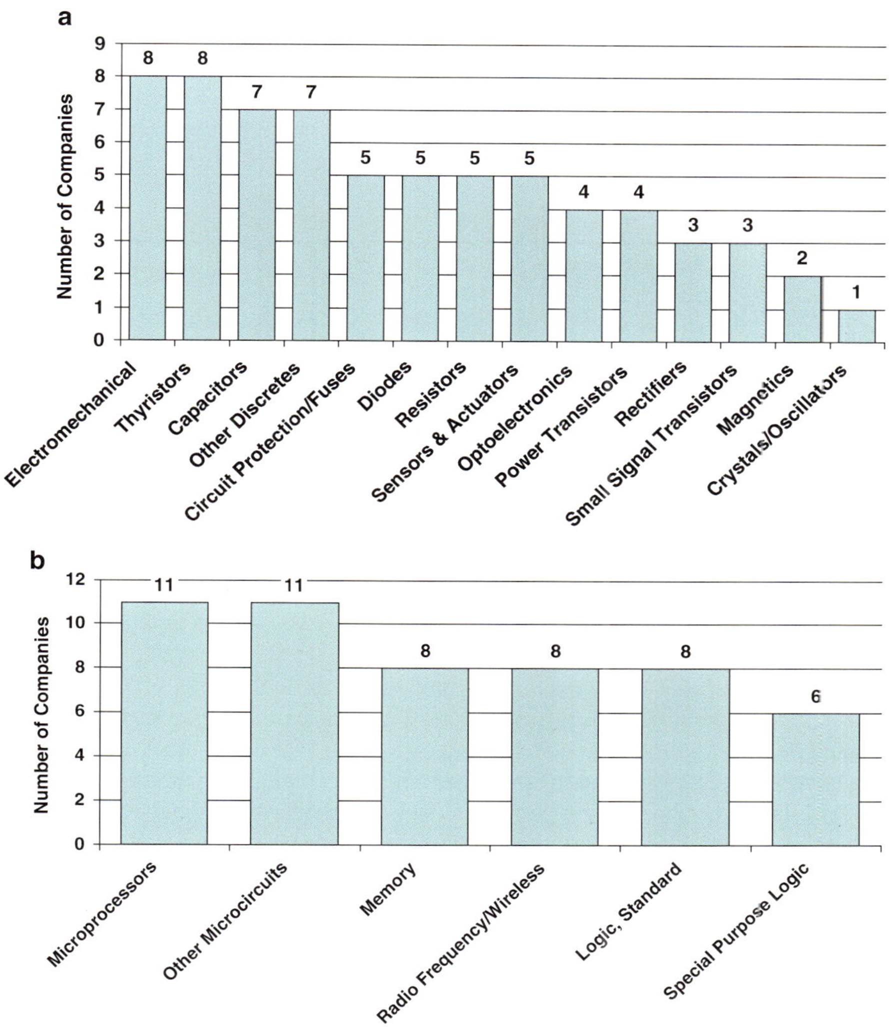

Fig. 1.4 Types of manufactured parts suspected/confirmed to be counterfeit. (**a**) Discretes. (**b**) Microcircuits

copies of the original designs" category (in Fig. 1.6a). A majority of counterfeit microcircuits were used and then remarked to a higher grade. Newly remarked microcircuits also contributed significantly to the counterfeit trade in discrete components. As with discrete components, fake and non-functional microcircuits were also reported (Fig. 1.6b).

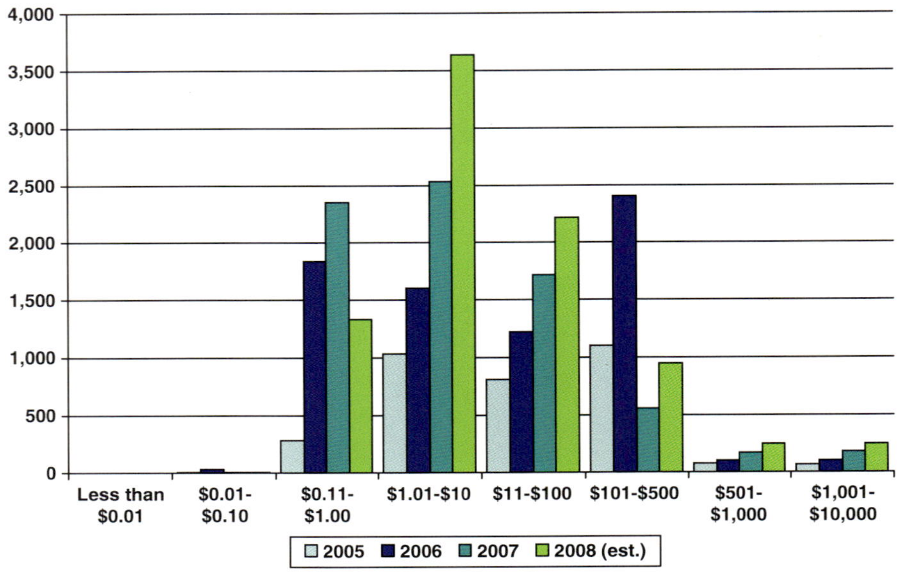

Fig. 1.5 Counterfeit incidents by product resale value—OCMs (2005–2008) [12]

Figure 1.7 shows how counterfeit components enter the supply chain. OCMs encountered counterfeit parts from at least 12 separate entities responsible for selling and distributing these parts. Brokers, independent distributors, and internet-exclusive suppliers were the main entities sourcing counterfeit parts. However, the OCMs also came across these parts from authorized distributors and even from US federal agencies.

1.5 Summary

This chapter presented a general overview of the practice of counterfeiting and counterfeit products. The reports discussed in this chapter showed that a wide variety of products ranging from luxury goods to electronic products or parts are prone to counterfeiting. With the advent of globalization, the scope of counterfeiting has encompassed countries and has become a global issue.

In terms of electronic components, counterfeiting has become rampant, spreading across the global electronics market. Counterfeit electronic products, a significant portion of the total counterfeit trade, can negatively impact not only the security and reliability of our critical systems, but also the research and development of these products. This chapter reported concrete data in order to present the grave issue of counterfeit electronic components and also touched upon the possible impacts on the economy, safety, and security of society at large.

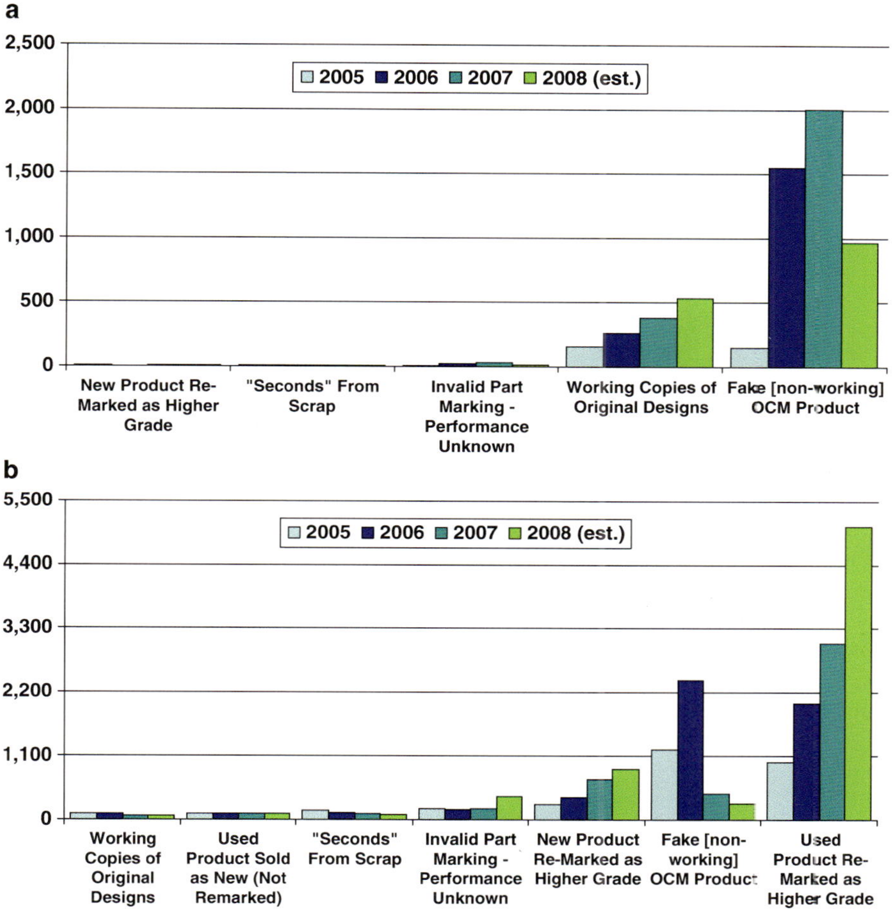

Fig. 1.6 Counterfeit incidents by type of problem. (**a**) Discretes. (**b**) Microcircuits

In the following chapters of this book, we will introduce all the different types of counterfeit electronic components, their detection and avoidance in the component supply chain, and the key challenges that must still be addressed to overcome the longstanding problem of electronics counterfeiting.

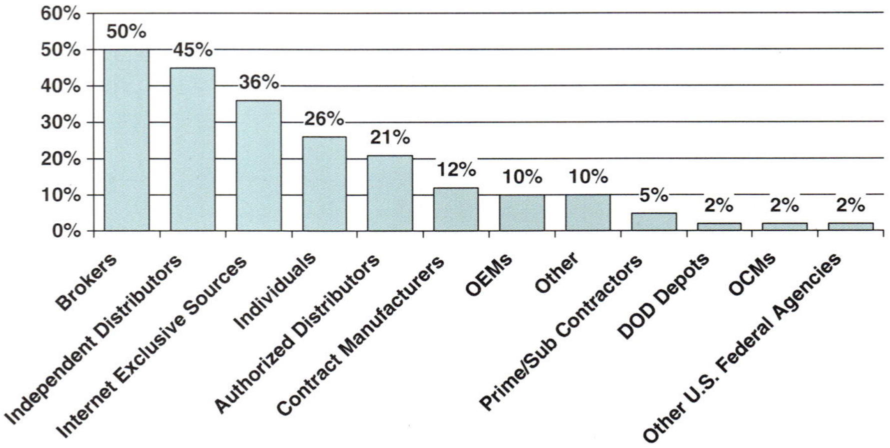

Fig. 1.7 Percent of OCMs with cases of counterfeit incidents sold by type of entity [12]

References

1. WHO, Counterfeit medicines: an update on estimates (November 2006), http://www.who.int/medicines/services/counterfeit/impact/TheNewEstimatesCounterfeit.pdf
2. P.E. Chaudhry, A. Zimmerman, *The Economics of Counterfeit Trade: Governments, Consumers, Pirates and Intellectual Property Rights* (Springer, Heidelberg, 2009)
3. D. Lague, Next step for counterfeiters: faking the whole company. The New York Times (May 2006)
4. The Federal Bureau of Investigation, Departments of justice and homeland security announce 30 convictions, more than $143 million in seizures from initiative targeting traffickers in counterfeit network hardware (May 2010)
5. A. Vance, Chasing pirates: inside Microsoft's war room. The New York Times (November 2010)
6. OECD, The economic impact of counterfeiting (1998)
7. OECD, The economic impact of counterfeiting and piracy: executive summary (2007)
8. OECD, Magnitude of counterfeiting and piracy of tangible products: an update (November 2009)
9. BASCAP, Estimating the global economic and social impacts of counterfeiting and piracy (February 2011)
10. U.S. Department of Homeland Security, Intellectual property rights seizures statistics: fiscal year 2013 (March 2014), http://www.cbp.gov/sites/default/files/documents/2013%20IPR%20Stats.pdf
11. U.S. Environmental Protection Agency, Electronic waste management in the united states through 2009 (May 2011)
12. U.S. Department Of Commerce, Defense industrial base assessment: counterfeit electronics (January 2010)

Chapter 2
Counterfeit Integrated Circuits

Counterfeit integrated circuits (ICs) pose a major concern to the industry and government as they potentially impact the security and reliability of a wide variety of electronic systems. A recent report [1] from the Information Handling Services Inc. [2] shows that reports of counterfeit parts have quadrupled since 2009 (see Fig. 2.1). This data has been compiled from two reporting entities—The Electronic Resellers Association International (ERAI) Inc. [3] and the Government-Industry Data Exchange Program (GIDEP) [4]. This report states that the majority of counterfeit incidents were reported by US-based military bodies and electronic firms from the aerospace industry.

Over the past couple of years, numerous reports [5] have pointed to counterfeiting issues in the US electronics component supply chain. One particularly prominent example of this problem is dramatized by the incident of Stephanie McCloskey, an administrator at VisionTech Components, LLC, who was sentenced to 38 months in prison for selling counterfeit ICs to the U.S. military and other crucial industries [6]. In November of 2010, McCloskey pled guilty to a federal charge of conspiracy to traffic counterfeit goods and mail fraud. Between 2006 and 2010, McCloskey conspired with Shannon L. Wren, the late owner of VisionTech Components, LLC to acquire counterfeit devices from China and Hong Kong, import them into various ports across the US, and market them on the VisionTech website as name-brand, trademark-protected ICs. From January 1, 2007 through December 31, 2009, Wren, McCloskey, and others generated nearly $15.8 million in gross receipts from the sale of their counterfeit ICs. The McCloskey conviction marked the first time anyone had been sentenced in a federal courtroom for trafficking ICs.

Another case in point of IC counterfeiting is Peter Picone, a 41-year-old man from Methuen, Massachusetts, who pled guilty in 2014 to importing thousands of counterfeit integrated circuits (ICs) from China and Hong Kong in order to resell them to US customers [7]. What transforms this case from the banal to the insidious is the fact that Picone targeted not only private consumers but also contractors who supplied these counterfeit ICs to the US Navy for use in nuclear submarines.

M. Tehranipoor et al., *Counterfeit Integrated Circuits*,
DOI 10.1007/978-3-319-11824-6_2

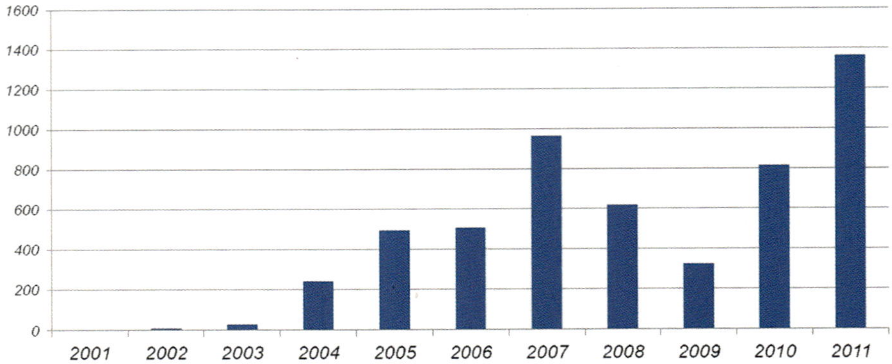

Fig. 2.1 Counterfeit incidents reported by IHS [1]

The contractors supplying the Navy specifically requested ICs that were new and not manufactured in China. Picone assured these contractors that the ICs were new and manufactured in Europe. However, tests conducted by the Navy have since revealed that the ICs purchased from Picone had been resurfaced to change the date code and to affix counterfeit marks, all in order to hide the parts' true origin.

In 2011, the US Senate Armed Services Committee reported about a grave counterfeiting incident, where the ice detection module on a new P-8A Poseidon aircraft was found to have counterfeit IC components [8, 9]. The ice detection system on an aircraft is a critical module that alerts pilots about the presence of ice on an aircraft's control surface, a potentially life-threatening situation. The module was found with an FPGA unit that had literally fallen out of its sockets and was found inside the module. On further inspection, it was found that the Xilinx FPGA component, which was badly used, worn-out, and was a discontinued model, was sold to BAE Systems, the P-8 aircraft component contractor, by Tandex Test Labs in California. Tandex had bought the component through an independent distributor that had acquired the part through a manufacturer in Shenzhen, China. Further investigation revealed that the module was reworked and had found its way through the supply chain to the P-8A aircraft.

In separate investigations, counterfeit IC components were also found on several essential military systems deployed by the US Army, such as high-altitude missiles, helicopters (SH-60B, AH-64 and CH-46), and aircrafts (C-17, C-130J and C-27J) [10]. In response to these incidents, Army Lt. Gen. Patrick J. O'Reilly of the Missile Defense Agency was quoted as saying, "We do not want to be in a position where the reliability of a $12 million THAAD interceptor is destroyed by a $2 part" [11].

In another incident in 2011, the display units onboard US military aircrafts were found with counterfeit electronic components [12]. The units, manufactured by L-3 Communication Display System, were meant for pilots to diagnose critical data such as engine fuel, location, and warning messages. Although any possible disasters were averted, a thorough investigation was conducted by L-3 and the

Senate Armed Services Committee, which traced the counterfeit components to Hong Dark Electronic Trade in Shenzen, China. Additional counterfeit components were found in other pieces of equipment onboard at least seven aircrafts, all were sold to the US Army by Raytheon and Boeing.

Unquestionably, the presence of such inferior and untested ICs in US Navy nuclear submarines could have catastrophic and far-reaching consequences. Because counterfeit ICs are vulnerable to unpredictable failures, they can lead to property damage, costly repairs, and bodily injury—even in the most benign of circumstances. However, when counterfeit ICs are used in systems of national importance, they also raise several national security concerns. Counterfeit ICs' histories are unknown, so it becomes unclear who has had access to them and how they have been altered. Such devices can be changed so that they contain malicious code or hidden "backdoors" that can disable systems, intercept communications, and intrude into computer networks. All of these issues imply massive consequences when placed in the context of US national security. In an attempt to tackle this counterfeiting epidemic, a Senate Armed Services public hearing and its later report clearly identified counterfeiting as a major issue to address [13, 14].

2.1 Counterfeit IC Types

With counterfeit incidents on the rise, it is increasingly important to understand what ICs are most likely counterfeit and what industries are impacted the most. Table 2.1 shows the five most commonly counterfeited components that represent $169 billion in potential annual risk for the global electronics supply chain. The components are as follows: analog ICs, microprocessor ICs, memory ICs, programmable logic ICs, and transistors. Together, these five types of components make up around 68 % (or, slightly more than two-thirds) of all the counterfeit incidents reported in 2011.

Table 2.2 shows the industries where these top five components are used. They include computing, consumer electronics, wireless and wired communications, automotive and industrial sectors. Automotive and industrial sectors involve critical systems and, thus, the appearance of unreliable counterfeit components in these applications is quite alarming. Untrustworthy counterfeit components are also a

Table 2.1 Top-five most counterfeited semiconductors in 2011 (percentage of counterfeit part reports) [15]

Rank	Commodity type	% of reported incidents
#1	Analog IC	25.2
#2	Microprocessor IC	13.4
#3	Memory IC	13.1
#4	Programmable logic IC	8.3
#5	Transistor	7.6
#6	Others	32.4

Source: IHS parts management 2012 [15]

Table 2.2 Percentage of market revenue for most commonly counterfeited product types by application market in 2011 (percentage share of revenue in millions of U.S. dollars) [15]

Part type	Industrial (%)	Automotive (%)	Consumer (%)	Wireless (%)	Wired (%)	Compute (%)	Other (%)
Analog IC	14	17	21	29	6	14	0
Microprocessor IC	4	1	4	2	3	85	0
Memory IC	3	2	13	26	2	53	1
Programmable logic IC	30	3	14	18	25	11	0
Transistor	22	12	25	8	10	22	0

concern for consumer applications where we are increasing becoming more reliant on electronic devices for computing, communication, online banking, handling personal data, etc.

In the following, we will present a taxonomy of counterfeit components. We will then present the vulnerabilities which lay in the different stages of electronic component supply chain and give rise to each counterfeit type. Finally, we will give a brief overview of the current state-of-the-art in the detection and avoidance of counterfeit electronic components.

2.2 Taxonomy of Counterfeit Types

The US Department of Commerce define a counterfeit component as one that

1. is an unauthorized copy;
2. does not conform to original OCM design, model, and/or performance standards;
3. is not produced by the OCM or is produced by unauthorized contractors;
4. is an off-specification, defective, or used OCM product sold as "new" or working; or
5. has incorrect or false markings and/or documentation.

The above definition does not include all possible scenarios where an entity in the component supply chain source electronic components that are authentic and certified by the OCMs. For example, one may copy the entire design of a component by reverse engineering [16, 17], manufacture them, and then sell them in the market under the OCM's identity. An untrusted foundry or assembly may source extra components without disclosing it to the OCMs [18, 19]. An adversary can insert a hardware Trojan [20] into a component to interrupt its normal operation and satisfy his/her own malevolent interests. All these scenarios impact the security and reliability of a system utilizing such components. Thus, we have expanded on the above definition of counterfeiting and developed a more comprehensive taxonomy of counterfeit types [21–26]. Figure 2.2 shows this novel taxonomy of counterfeit types. Descriptions of each type are given in the subsections below.

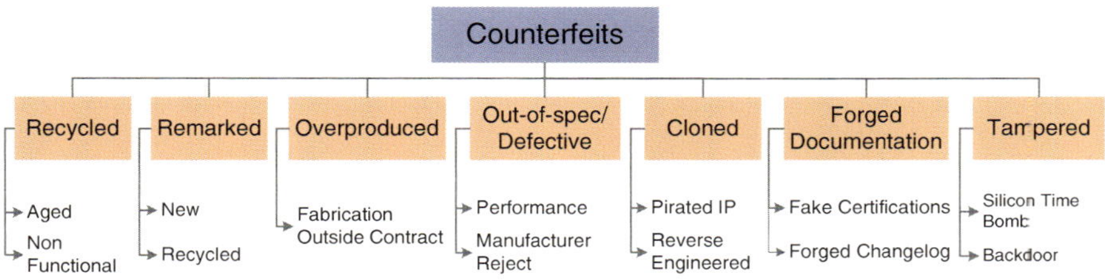

Fig. 2.2 Taxonomy of counterfeit types

2.2.1 Recycled

The term "recycled" refers to an electronic component that is reclaimed or recovered from a system, and is then modified to be misrepresented as a new component of an OCM. Recycled parts may exhibit lower performance and have a shorter lifetime due to aging from their prior usage. Further, the reclaiming process (removal under a very high temperature, aggressive physical removal from boards, washing, sanding, repackaging, etc.) could damage the part(s), introduce latent defects that pass initial testing but are prone to failure in later stages, or make them completely non-functional due to exposure to extreme conditions in an uncontrolled environment. Such parts will, of course, be unreliable and render the systems that unknowingly incorporate them equally unreliable.

The United States Committee on Armed Services held a hearing regarding an investigation of counterfeit electronic parts in the defense supply chain and the investigation revealed that e-waste from discarded electronic components are being used for these recycled counterfeit parts [13, 14]. In the United States, only 25 % of electronic waste was properly recycled in 2009 [28]. These figures might be comparable or even worse for many other countries. This huge resource of e-waste allows counterfeiters to pile up an extremely large supply of components. These components are then recycled from the stockpile of e-waste using a crude process. A typical recycling process is as follows [29]:

1. The recycler collects discarded printed circuit boards (PCBs) from which used components (digital ICs, analog ICs, capacitors, resistors, etc.) can be harvested.
2. The PCBs are heated over an oven flame. When the soldering material starts to melt, the recycler smashes the PCB over a bucket to detach and collect the components.
3. The original marking of the components are removed by microblasting where blasting agents are bombarded on a component's surface. Compressed air is generally used to accelerate the blasting particles. Some popular blasting agents include aluminum oxide powder, sodium bicarbonate powder, and glass bead. The choice of blasting agent depends on the components' package type such as dual in-line package (DIP), plastic leaded chip carrier (PLCC), etc.

4. A new coating material is applied to the component by using black topping and resurfacing.
5. New markings containing a PIN number, date/lot code, manufacturer logo, country of manufacture, etc., are then printed either by ink printing or laser printing on the new black topped surface.
6. The component leads, balls and/or columns are reworked (cleaning and straightening of leads, replating leads with new materials, forming new solder balls, etc.) to make them appear new.

Figure 2.3 shows a recycling process documented by NASA [27]. Clearly, the recycling process impacts the reliability of recycled components as they are subjected to harsh handling practices and impacts such as the following:

1. The components are not protected against electrostatic discharge (ESD) and electrical overstress (EOS).
2. Moisture sensitive components are not properly baked and dry-packed.
3. The components may be damaged due to (a) high recycling temperature, (b) mechanical shock due to smashing and other handling, (c) humidity levels from cleaning with water and storage in damp conditions, and (d) other mechanical and environmental stress resulting from the recycling process.

In effect, the recycled components are degraded even further by such processes. This only exacerbates the prior effects of aging due to usage of the component in a system.

Fig. 2.3 A typical IC recycling process [27]

The recycled components are discussed widely by the government, industry and test labs. The standards [30–33] recommends different test plan to detect these components. In this book, our aim is to highlight the most effective ways of detecting these components. Chapters 3–8 describe various test methods to detect these components. Chapter 9 exclusively describes design-for-anti-counterfeit (DFAC) measures to easily detect recycled components and therefore prevent them from getting into the component supply chain.

2.2.2 Remarked

Electronic components contain markings on their packages to uniquely identify them and their functionality. The marking contains information such as the part or identifying number (PIN), lot identification code or date code, device manufacturer's identification, country of manufacture, electrostatic discharge (ESD) sensitivity identifier, certification mark, and so forth. A detailed description of markings can be found in Section 3.9.5 of MIL-PRF-38534H [34].

Clearly, a component's markings are very important. They identify component's origin and, most importantly, determine how the component should be handled and used. For example, a space grade component can withstand a wide range of temperatures, radiation levels, etc. that would cause instant failure for a commercial grade component. The component manufacturer, grade, etc. also determine how much the component is worth. The price of space and military grade components can be phenomenally higher than commercial grade components. For example, a BAE radiation-hardened processor such as the RAD750 could cost in the range of $200,000, compared to a commercial processor which could be in the range of a few hundred dollars [35]. These space-grade processors are used in satellites, rovers and space shuttles, and are designed to withstand a wide range of temperatures and radiation levels typically found in space. Herein lies the incentive for remarking a component (i.e., changing its original markings) as well as the threat created by remarking. A counterfeiter can drive up a component's price on the open market by changing its markings to that of a higher grade or better manufacturer. However, such remarked components will not be able to withstand the harsh conditions of their more durable, higher-grade counterparts. This can create substantial issues if such components end up in critical systems.

A notable example of this is the P-8A Poseidon Aircraft incident, which was brought to light during a hearing held by the US Senate Committee on Armed Services in 2011 [36]. It was found that the ice-detection module aboard the P-8A Poseidon aircrafts, which transports anti-submarine and anti-surface warfare missiles, was found with counterfeit FPGA units. The ice-detection module is a critical component which warns a pilot of ice that has developed on the surface of the aircraft. In this case, it was found that the FPGA units controlling the module were used and falsely remarked as being produced by Xilinx. On further analysis of the supply chain, the components were actually traced back to a manufacturer in Shenzhen, China.

It is fairly easy to remark a component that is indistinguishable from the original markings to the naked eye. A component is first prepared for remarking by either chemically or physically removing the original marking and then by blacktopping (resurfacing) the surface to hide any physical marks or imperfections that have been left from the marking removal process. False markings are then printed either by laser marking or ink marking onto the components to appear as though produced by the OCMs. Ink marking can be performed in various ways. The fast and flexible ink jet printing is a popular choice for generating ink marking. Some other methods include stamping, screen printing, transfer printing, and pad printing [37]. Similarly, laser marking is also very flexible and generally etched by CO_2 or YAG laser [37].

Similar to the recycled counterfeit type, remarked components are also extensively discussed by the government, industry and test labs. The standards developed thus far recommended the same test plan to detect remarked components. In this book, we address the detection of recycled and remarked components simultaneously, which is described in detail in Chaps. 3–8.

2.2.3 Overproduced

Today's high-density integrated circuits are mostly manufactured in state-of-the-art fabrication facilities. Building or maintaining such facilities for modern CMOS technologies is reported to cost more than several billions of dollars and this number is growing with each new technology node [38]. Given this increasing cost and the complexity of foundries and their processes, the semiconductor business has largely shifted to a contract foundry business model (horizontal business model) over the past two decades.

Figure 2.4 shows the trust and security issues due to the practice of horizontal model. In this model, the design houses outsource their designs for fabrication and packaging to companies all around the world, mainly to reduce manufacturing costs. Although the contracted parties may agree to only manufacture a certain number of working components, they could in fact exceed this amount. Untrusted foundries and assemblies may produce more than the number of components they are contracted to produce. In addition, they can overbuild components by hiding the actual yield (i.e. the percentage of defect-free components to the total number of components) information. Thus, the foundries and assemblies overproduce so that the required number of components can be met. Unfortunately, since the fabrication and assembly process are often performed by third parties, most of which are overseas, the design houses cannot monitor the fabrication and assembly process nor get the actual yield information. Here, the likely risk is that, these components that are in excess of the contracted amount could be sold by the foundries and assemblies without any knowledge of the design houses.

This process of manufacturing and selling outside of the agreement with the design house (i.e., the components' intellectual property (IP) owner) is known as "overproduction". A well understood concern with overproduction is the inevitable

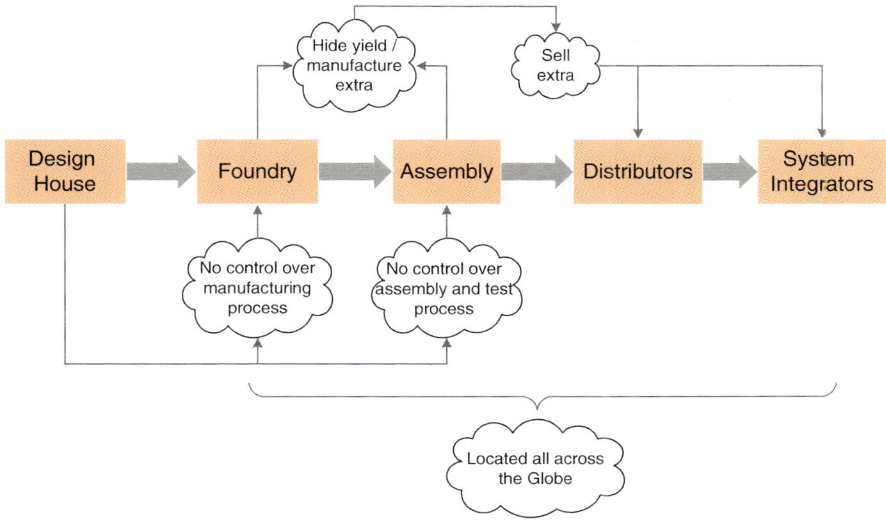

Fig. 2.4 Trust and security issues due to globalization

loss in profit for the design houses. Design companies usually invest a large amount of time and effort into research and development (R&D) of their products. When an untrusted foundry or assembly overproduces and sells these components, the design house loses any possible revenue that could have been gained from those components. However, an even bigger concern with overproduced components is that of reliability. Overproduced components may simply end up in the market with minimal or no testing for reliability and functionality. These components may find their way back into the supply chain for many critical applications such as military equipment and consumer products, which raises concern for safety and reliability. Further, since these components bear the same name of the design houses, failure of these components would then tarnish the reputation of the original component manufacturer. In Chap. 11 we will discuss overproduction in detail and will describe secure split test to detect these overproduced ICs.

2.2.4 Out-of-Spec/Defective

A part is considered defective if it produces an incorrect response in post-manufacturing tests. The manufacturing tests are performed in multiple stages. Figure 2.5 shows a typical manufacturing test process [39]. During the manufacturing process, the first test performed is the wafer test to inspect which ICs, fabricated on the wafer, are defective. If there are too many defective ICs on the wafer, the foundry sometimes rejects the whole wafer. A wafer generally contains

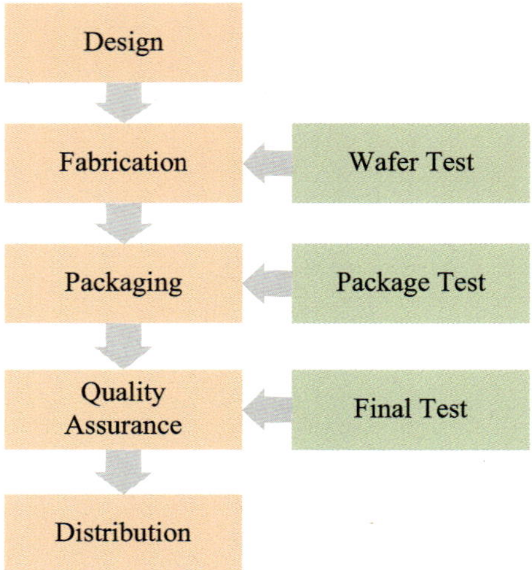

Fig. 2.5 Manufacturing test process

hundreds of ICs depending on the size and type of ICs and may worth hundreds of dollars. An untrusted entity may source these defective wafers to an assembly and produce defective or out-of-specification ICs.

After wafer tests, the defect free chips are sent to assembly for packaging. The healthy chips are then sorted out by using package tests and the chips that have been damaged during the packaging process are discarded. An untrusted entity again can supply these chips into the supply chain. The final test is performed as a part of quality assurance of the final packaged chips before sending them to the market. Burn-in, using accelerated temperature and voltage, is often performed to test latent defects in order to avoid the failures in the early operational stages of chips.

All the rejected chips from various test process should be destroyed (if they are non-functional), downgraded (does not satisfy the specification), or otherwise be properly disposed of. However, if they are sold on the open market instead, either knowingly by an untrusted entity or by a third party who has stolen them, there will be an inevitable increase in their risk of failure.

The detection of these defective/out-of-spec components is not an easy task. It may be easy to detect a defective chip that has been rejected in the early test process by using simple parametric tests. However, it will be extremely difficult if the chips are rejected in the later phase of the test process. The entity in the test process must be completely familiar with the internal structure and functionality of the design, which often is not the case. The only way we can truly prevent those chips from getting into the supply chain by placing DFAC measures during the design phase of those chips. We will describe such a DFAC measure, secure split test, in Chap. 11 to prevent these defective chips into the market.

2.2.5 Cloned

Cloning is widely used by a range of adversaries/counterfeiters (from small entities to large organizations) to copy a design in order to eliminate the large research and development (R&D) costs of a part. Cloning is a major concern for semiconductor intellectual property (IP), such as layouts, netlists and HDL design blocks (refer to Chap. 10 for a detailed discussion on semiconductor IP types), as well as fabricated integrated circuits. Cloning can be done by reverse engineering or by illegally obtaining semiconductor intellectual property (IP) such as layouts, netlists etc. (also called IP theft).

Reverse engineering (RE) [16, 17] is the process of examining an original component in order to fully understand its nature and functionality. It can be achieved by extracting the physical interconnection information layer-by-layer destructively or non-destructively followed by image processing analysis to reconstruct the complete structure for a component [16, 40]. The prime motivation for reverse engineering a component is to make an existing copy of it often by the competitors of the OCM. An entity involved in reverse engineering often possesses expensive and sophisticated instruments. Scanning electron microscope (SEM) or transmission electron microscope (TEM) are commonly used to take images of each layer of a component after delayering. An automated software can be used to stitch the images together to form a complete structure. For example, ICWorks Extractor from Chipworks Inc. (Ottawa, Canada) has the capability to form a 3D structure by combining all the images from the internal layers of a chip [16].

Cloning can also occur by unauthorized knowledge transfer from a person with access to the part's design. In order to provide proof of authorship, watermarks in various forms such as power signatures, design constraints etc. are added to semiconductor IPs (refer to Chap. 10 for IP watermarking techniques). If watermarking strategies are not implemented or are weak, it may be possible for counterfeiters or personnel possessing unauthorized knowledge of the IP to simply copy the IP, make cloned semiconductor components and market them for profit. Such cloned components violate intellectual property rights of the rightful IP owners and could cause them significant losses in revenue. We will also present a novel DFAC measure to prevent cloned ICs getting into the supply chain in Chap. 11.

2.2.6 Forged Documentation

The documentation shipped with any component contains information regarding its specifications, testing, certificates of conformance (CoC) and statement of work (SoW). By modifying or forging these documents, a component can be misrepresented and sold even if it is nonconforming or defective. It is often difficult to verify the authenticity of such documents because the archived information for older designs and older parts may not be available at the OCM. Legitimate documentation can also be copied and associated with parts from a lot that do not correspond with the legitimate documentation.

The incentive for counterfeiters and risks associated with parts with forged documentation are similar to those discussed above for remarking.

2.2.7 Tampered

The vulnerabilities of integrated circuits to malicious alteration has become predominant due to the globalization of the semiconductor supply chain (see Fig. 2.4). ICs that have been tampered with can have dangerous consequences to military infrastructures, aerospace systems, medical, financial, and transportation infrastructures, and commercial infrastructures.

An adversary can insert a hardware Trojan [20] in a design to interrupt its normal operation and/or disable it in the future, effectively making it a "silicon time bomb". A hardware Trojan may also create a backdoor that gives access to critical system functionality or leaks secret information to an adversary. Hardware Trojans can be implemented by modifying (1) the hardware in application-specific integrated circuits (ASICs), digital signal processors (DSPs), microprocessors, microcontrollers, or (2) the bitstream for field programmable gate arrays (FPGAs). A hardware Trojan can modify the functionality of a design in a variety of ways. For example, a hardware Trojan can disable the crypto module on a design and leak unencrypted plain text which can easily be intercepted, or disable the system clock of a module for a small duration to launch a sabotage. A detailed taxonomy for hardware Trojan can be found in [41].

Along with hardware Trojans, tampering can be performed after the fabrication of parts by circuit editing [42]. With the advent of nanoscale manipulation technologies such as focused ion beam (FIB), it has been reported that adversaries can modify the circuit netlist even in the linewidths of 20 nm and pitches of 40 nm range [43]. In this approach, the adversary can cut wires that connect transistors/gates, or create a connection by rerouting between the transistors/gates to modify the functionality of a circuit.

Since the detection and avoidance of tampering is a large problem unto itself, we shall consider it beyond the scope of this book. Interested readers are suggested to read "Integrated Circuit Authentication: Hardware Trojans and Counterfeit Detection" [20] for further information on hardware Trojan insertion, detection, and prevention.

2.3 Supply Chain Vulnerabilities

Typically an electronic component will go through a process like the one shown in Fig. 2.6. This process includes design, fabrication, assembly, distribution, usage in the system, and, finally, end of life (resign). The vulnerabilities associated with each step are discussed in more details below.

2.3.1 *Design*

The design implementation of large complex integrated circuits has evolved to a stage where it is extremely challenging to complete the entire design in-house. The flow from RTL to GDSII is performed in many different places (even in different countries) mainly to reduce development costs and design-to-market time (see Fig. 2.4). Design reuse has also become an integral part of SoC design, where hard IPs (layout level designs), firm IPs (netlists and HDL designs with parameterized constraints), and soft IPs (synthesizable register-transfer level (RTL) designs) are reused by designers (more specifically, system integrators who combine various IPs to create new IC designs) for simplifying complex design problems. Attacks on the design stage can be performed in the following two ways: (1) one or more rogue employees can steal these IPs used in the system and pirate them to another entity, which has the manufacturing capability to fabricate cloned components, or (2) the design uses an IP which is deliberately tampered with malicious codes to modify functionality and/or create backdoors to leak secret information.

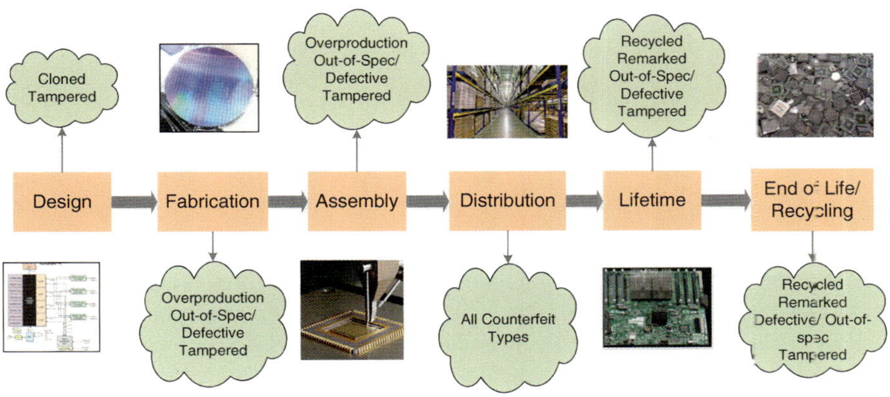

Fig. 2.6 Vulnerabilities in the electronic components supply chain

2.3.2 Fabrication

As discussed above, today's integrated circuits are manufactured in fabrication facilities (fabs) located all around the world due to the high costs associated with owning and maintaining new fabs. Design houses go into contract with foundries to fabricate their designs, the design houses disclose the details of their IPs, and they also pay for mask-building costs based on their designs. The contract agreement between the foundry and design house is protected by IP rights [44]. However, this horizontal business model creates a trust issue between the design house and the foundry. An untrusted foundry can potentially (1) make extra/overproduced ICs, by hiding their yield, and selling those extra ICs in the open market, (2) clone the design, and (3) source defective and out-of-specification wafers or dies to packaging companies to make finished parts.

2.3.3 Assembly

After fabrication, the foundry sends tested wafers to the assembly line to cut the wafers into die, package the die, and perform final tests before being shipped to the market. Figure 2.5 shows the test process where the package test is performed at the assembly, and then the final test for the quality assurance. An untrusted assembly may supply defective and out-of-spec chips rejected by these test processes to the market. These assemblies can also hide the yield information of their packaging process like untrusted foundries, and can stockpile those extra components. An untrusted assembly may also repackage recycled dies, remark them as new, and may also upgrade a component by printing higher grade information on the part package.

2.3.4 Distribution

The tested ICs are sent either to the distributors or to system integrators. The distributors sell these ICs in the market. These distributors are of several types—OCM authorized distributors, independent distributors, internet-exclusive suppliers, brokers etc. The most significant threat lies with those outside of OCM-authorized distributors that are not officially associated with or contracted to OCMs. A recent report [45] from the Semiconductor Industry Association pointed out that counterfeit components could potentially be avoided by exclusively buying these components either directly from the OCM or directly from the OCM authorized distributors or resellers. However, it is also worth noting that OCMs have come across counterfeit parts from OCM-authorized distributors and even from US federal agencies [46].

2.3.5 *System Integration/Lifetime*

System integration is the process of combining together all the components and subsystems into one complete system. An untrusted system integrator can potentially use all types of counterfeit components in their system. They can maximize their profit by using cheap or tampered counterfeit components to drive their costs down and potentially inflate the actual worth of the final system.

2.3.6 *End-of-Life*

When electronics age or become outdated, they are typically retired or resigned and then subsequently replaced. Proper disposal techniques are highly advised to extract precious metals and to prevent hazardous materials (lead, chromium, mercury, etc.) from harming the environment [47]. Yet, these techniques are largely ignored, resulting in a large amount of electronic waste or e-waste. For instance, in the United States, only 25 % of electronic waste was properly recycled in 2009 [28]. As discussed in Sect. 2.2.1, a profitable business has grown out of reclaiming used components from this e-waste, remarking them, and then re-inserting them into the supply chain as new components. According to current reports, these recycled and remarked components account for over 80 % of the reported counterfeit parts in the supply chain [48] and represent a growing threat [49]. Also, at this stage, the counterfeiters can potentially tamper with used components for the purposes of sabotage or malfunction.

2.4 Detection and Avoidance of Counterfeit ICs

The detection and avoidance of counterfeit parts is a multifaceted problem, and we believe that it is still in its infancy. Currently, efforts at detection have mainly focused on identifying counterfeit parts that are already circulating in the electronic component supply chain by performing a sequence of "post-counterfeit" detection methods. On the other hand, avoidance measures (also termed as design-for-anti-counterfeit (DFAC) measures) can help to prevent these counterfeit parts from entering into supply chain in the first place. Detection could then be easily performed by observing the response of these DFAC measures.

While developing the plans for detection and avoidance of counterfeit components, it is necessary to consider all different components, which can be classified by their type, size, and state (see Fig. 2.7). The type is classified into three categories, namely analog, digital, and mixed-signal, depending on the applications they are used. The components can be of different sizes, namely large, medium, and small, depending on the die. It is also imperative that the larger components have more

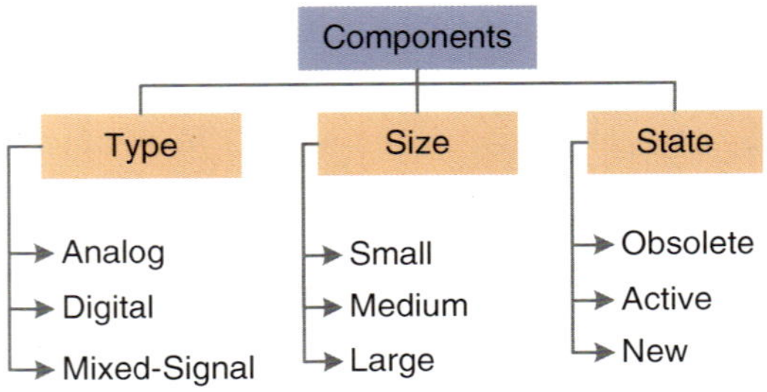

Fig. 2.7 Taxonomy of component types

input/output (IO) pin counts. We categorize state into three distinct types—obsolete, active, and new. Obsolete refers to components that are no longer manufactured by the original component manufacturers (OCM), as they may have switched to newer designs or new technology nodes to improve performance, reliability, and/or manufacturing cost. These components would only be available through OCM authorized or independent distributors of electronic components. Active components are still being manufactured by OCMs, but their designs cannot be changed because of—(1) the extra cost of developing new masks and (2) performance and reliability concerns. Thus, both obsolete and active components present very little opportunities for integration of anti-counterfeiting measures. New components are very flexible in implementing avoidance measures as they are in the design phase where the OCM can still modify masks and validate performance.

2.4.1 Current Status of Detection

Different types of counterfeit components pose unique challenges to develop a test plan for the detection of counterfeit components. As obsolete parts are no longer being manufactured, and active parts are being fabricated based on previous designs and developed masks, the focus should be on the "detection" of such counterfeit components. On the other hand, for new components it is possible to integrate anti-counterfeit strategies during the design process that may prevent counterfeiting altogether or facilitate more accurate and efficient detection.

There are a few standards available today to guide users in detecting counterfeit parts. The group responsible for many of these standards is the G-19 Counterfeit Electronic Parts Committee, set forth by SAE International [50]. Their standards target three different sectors of the industry: distributors, users, and test service providers (i.e., test laboratories). These standards are as follows:

1. AS5553—Counterfeit Electronic Parts; Avoidance, Detection, Mitigation, and Disposition [30];
2. ARP6178—Fraudulent/Counterfeit Electronic Parts; Tool for Risk Assessment of Distributors [51];

3. AS6081—Fraudulent/Counterfeit Electronic Parts: Avoidance, Detection, Mitigation, and Disposition—Distributors Counterfeit Electronic Parts; Avoidance Protocol, Distributors [52] (intended for independent distributors and brokers);
4. AS6496—Fraudulent/Counterfeit Electronic Parts: Avoidance, Detection, Mitigation, and Disposition—Authorized/Franchised Distribution [53]; and
5. AS6171—Test Methods Standard; Counterfeit Electronic Parts [31].

While SAE is the most prominent entity when it comes to standards, there are a couple of programs designed to help independent distributors gain users' trust. Components Technology Institute, Inc. (CTI) [54] has created the Counterfeit Components Avoidance Program (CCAP-101) [32]. Independent distributors can be certified as CCAP-101 compliant, which is accomplished by going through a yearly audit. Another program with similar goals has been developed by the Independent Distributors of Electronics Association (IDEA) [33].

All the above standards focus primarily on the evaluation of physical properties of components by using physical tests (see in Chap. 4) instead of evaluating of the electrical parameters by using electrical tests (see Chap. 5) to decide whether a part is a counterfeit or not. There are still many challenges and limitations associated with them. First, these tests are mostly designed for detecting recycled and remarked counterfeit types, and are less effective in detecting the rest (overproduced, cloned, etc.). Second, test time and cost are major bottlenecks for applying the tests. For instance, some of the physical tests require hours to inspect only a single component. Different test setups are necessary for functional verification by using electrical tests, which makes them extremely expensive as well. In addition to these, the destructive nature of physical tests implies that it may be impossible to adequately test all components. Finally, most of these physical tests are performed in ad-hoc fashion without automation. Lack of test metrics leaves the task of interpreting test results in the hands of subject matter experts.

These challenges and limitations are described in detail in Chaps. 4 and 5. In addition to them, another major issue with many standards is that the policy and the regulations are their main focus rather than technology. Therefore, it is easy for counterfeiters to adapt to the new regulations, thus circumventing the effective detection of counterfeit parts. However, we are very hopeful that the ongoing activity of developing "AS6171 - Test Methods Standard; Counterfeit Electronic Parts" [31] by G-19A group, will provide users with the necessary means to fight against counterfeiting.

We shall elaborate on the test methods in Chaps. 4 and 5. In Chap. 6, we will introduce test metrics to assess these tests. Also, we will present a comprehensive framework to select the best set of test methods to maximize the test confidence under test cost and time constraint.

2.4.2 *Current Status of Avoidance*

As discussed above, many of the physical and electrical tests in current standards are limited in scope, have large test time and cost, etc. It is thus necessary to implement design-for-anti-counterfeit measures for all types of components in order to prevent this widespread penetration of counterfeit components in the future. These measures helps us to detect counterfeit components without performing conventional physical and electrical tests. However, design of effective measures can be challenging for different types of components (shown in Fig. 2.7), which each have their own unique limitations. For new components, anti-counterfeit mechanisms can be integrated into the die of ICs during the design phase. For active and obsolete components, alternative DFAC measures could be placed on the packaging.

Figure 2.8 shows the current technologies available for the avoidance of counterfeit parts in the component supply chain. The x-axis and y-axis represent the counterfeit types and component types respectively. The component types in the

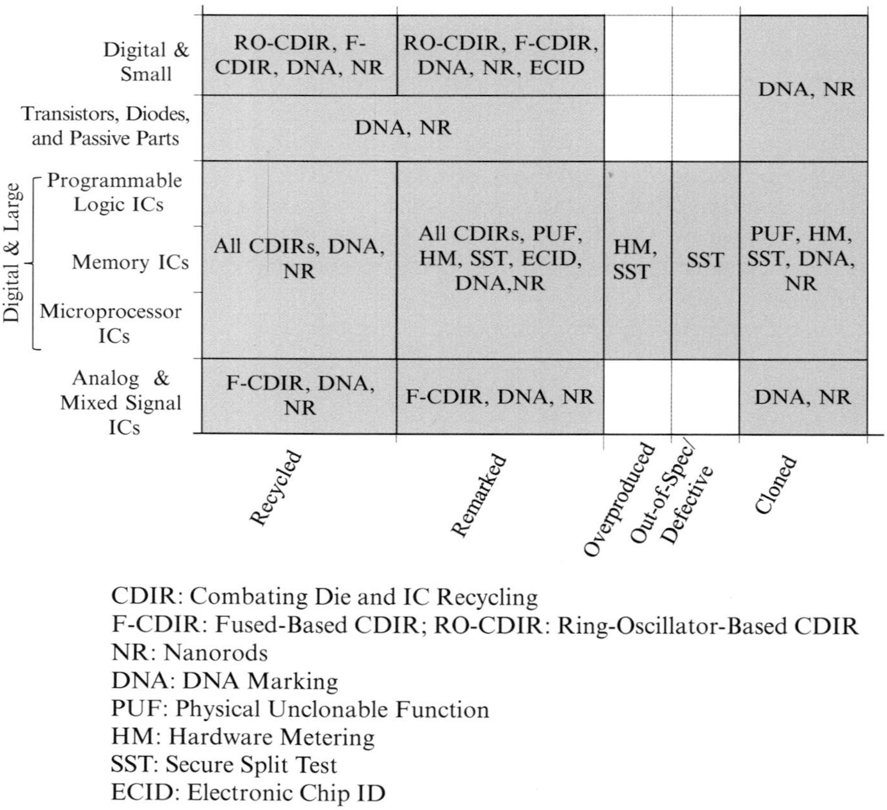

Fig. 2.8 Current technologies for the avoidance of counterfeit parts in the component supply chain

y-axis are arranged top to bottom from lowest to highest frequency of counterfeit incidents in the supply chain [15]. The sheer number of component types (analog, digital, and mixed-signal) and sizes (large or small) exists in the component supply chain. They are usually unique in their structures and functionalities. In addition, there exists seven different types of counterfeit components, which require unique measures for their detection. Thus, it extremely challenging to find a one-size-fits-all solution to detect counterfeit components and prevent them getting into the component supply chain. In Fig. 2.8, we designate different DFAC measures to different counterfeit types (shown in x-axis) and component types and sizes (shown in y-axis). We will present all these different DFAC measures in Chaps. 9–12. In the following, we provide a short description for these chapters.

Chapter 9 introduces several low-cost combating die and IC recycling (CDIR) structures. The ring-oscillator-based CDIR (RO-CDIR) structure can be implemented in digital ICs with new technology nodes, while the antifuse-based CDIR (AF-CDIR) structure can be placed in large digital ICs of new and older technology nodes. The low-cost fuse-based CDIR (F-CDIR) structure can be implemented in any components (small/large, analog/digital) and any technology node.

Physical unclonable functions (PUFs) [55], secure split test (SST) [19], hardware metering (HM) [18] and electronic chip ID (ECID) [56] can be implemented to detect large, remarked digital components. We briefly describe SST in Chap. 11 as it is the only technology available today that can potentially detect out-of-spec/defective counterfeit types. We will also describe PUFs and hardware metering in that chapter.

In Chap. 12, we introduce package based marking—DNA markings (DNA) [57] and nanorods (NR) [58]—that can potentially be implemented in all component types. These technologies can be implemented to detect cloned ICs along with recycled and remarked counterfeit types.

2.5 Summary

In this chapter, we presented a comprehensive overview of the issue of counterfeit integrated circuits. With the recent surge in counterfeit products, counterfeiting has pervaded the electronics supply chain. Various IC components such as processors, FPGAs and analog/mixed-signal/digital ICs have all been subject to counterfeiting. Keeping in mind that these ICs are used in automotive, healthcare, military and countless other critical infrastructures, the possible hazards caused by counterfeit ICs are startling. In addition, the illegal market of counterfeit ICs also affects the economy, causing significant losses in revenue.

We also presented a detailed discussion on the various types of counterfeit components, which include components that could be recycled, remarked, overproduced, defective/out-of-spec, cloned, tampered or have forged documentation. Each type of counterfeit component could appear at various stages in the electronic component supply chain, such as design, fabrication, assembly, distribution, system integration and end-of-life. For example, overproduced components are likely to appear during the fabrication process and recycled components could originate after components have reached their intended end-of-life. In order to combat the issue of counterfeit ICs, suitable detection and avoidance mechanisms are necessary. Detection of counterfeit components focuses on identifying counterfeits already circulating in the supply chain. We also introduced the various standards available, such as the G-19 Standard to guide the detection of counterfeit components. Nonetheless, detection mechanisms are prone to issues such as the large cost and time involved. More importantly, they address the issue of counterfeiting only after it has occurred. This brings up counterfeit avoidance mechanisms, which involve design-for-anti-counterfeit measures such as hardware metering (HM) and secure split test (SST) for preventing counterfeit ICs from reaching the supply chain in the first place. Unfortunately, since ICs could be of various types (analog, digital or mixed signal) and sizes, a one-solution-fits-all approach for detection is hard to implement. Chapters 9–12 will focus on measures for the detection and avoidance of these components.

References

1. J. Cassell, Reports of counterfeit parts quadruple since 2009, challenging US Defence Industry and National Security IHS Pressroom (April 2012), http://press.ihs.com/press-release/design-supply-chain/reports-counterfeit-parts-quadruple-2009-challenging-us-defense-in
2. Information Handling Services Inc. (IHS), http://www.ihs.com/
3. ERAI, Report to ERAI, http://www.erai.com/information_sharing_high_risk_parts
4. GIDEP, Government-Industry Data Exchange Program (GIDEP), http://www.gidep.org/
5. trust-HUB, http://trust-hub.org/home
6. U.S. Department of Justice, Administrator of VisionTech Components, LLC sentenced to 38 months in prison for her role in sales of counterfeit integrated circuits destined to U.S. military and other industries. Press Releases (October 2011), http://www.justice.gov/usao/dc/news/2011/oct/11-472.html
7. U.S. Department of Justice, Massachusetts man pleads guilty to importing and selling counterfeit intergrated circuits from China and Hong Kong (June 2014), http://www.justice.gov/opa/pr/2014/June/14-crm-595.html
8. B. Carey, Senate inquiry finds widespread counterfeit components. AIN (June 1 2012), http://www.ainonline.com/aviation-news/ain-defense-perspective/2012-06-01/senate-inquiry-finds-widespread-counterfeit-components
9. J. Reed, Counterfeit parts found on P-8 Posiedons. Defensetech (November 2011), http://defensetech.org/2011/11/08/counterfeit-parts-found-on-new-p-8-posiedons/
10. R. McCormack, Boeing's planes are riddled with chinese counterfeit electronic components. Manufacturing and Technology News **19** (June 2012)
11. T. Kaiser, SAS committee: counterfeit electronics from China could be harmful to military. Dailytech (November 2011)
12. T. Capaccio, China Counterfeit Parts in U.S. Military Aircraft. Bloomberg (November 2011)
13. U.S. Senate Committee on Armed Services, Inquiry into counterfeit electronic parts in the Department Of Defence Supply Chain (May 2012)
14. U.S. Senate Committee on Armed Services, Suspect counterfeit electronic parts can be found on internet purchasing platforms (February 2012), http://www.gao.gov/assets/590/588736.pdf

15. IHS iSuppli, Top 5 most counterfeited parts represent a $169 billion potential challenge for global semiconductor market (2011)
16. R. Torrance, D. James, The state-of-the-art in ic reverse engineering, in *Proceedings of the 11th International Workshop on Cryptographic Hardware and Embedded Systems*. CHES '09. (Springer, 2009, Berlin), pp. 363–381. http://dx.doi.org/10.1007/978-3-642-04138-9_26
17. I. McLoughlin, Secure embedded systems: the threat of reverse engineering, in *Parallel and Distributed Systems, 2008. ICPADS '08. 14th IEEE International Conference on* (December 2008), pp. 729–736
18. F. Koushanfar, G. Qu, Hardware metering, in *Proc. IEEE-ACM Design Automation Conference* (2001), pp. 490–493
19. G. Contreras, T. Rahman, M. Tehranipoor, Secure split-test for preventing IC piracy by untrusted foundry and assembly, in *Proc. International Symposium on Fault and Defect Tolerance in VLSI Systems* (2013)
20. M. Tehranipoor, H. Salmani, X. Zhang, *Integrated Circuit Authentication: Hardware Trojans and Counterfeit Detection* (Springer, Zurich, 2014)
21. U. Guin, D. DiMase, M. Tehranipoor, A comprehensive framework for counterfeit defect coverage analysis and detection assessment. J. Electron. Test. **30**(1), 25–40 (2014)
22. U. Guin, D. DiMase, M. Tehranipoor, Counterfeit integrated circuits: Detection, avoidance, and the challenges ahead. J. Electron. Test. **30**(1), 9–23 (2014)
23. U. Guin, K. Huang, D. DiMase, J. Carulli, M. Tehranipoor, Y. Makris, Counterfeit integrated circuits: a rising threat in the global semiconductor supply chain. Proc. IEEE **102**(8), 1207–1228 (2014)
24. U. Guin, D. Forte, M. Tehranipoor, Anti-counterfeit techniques: from design to resign, in *Microprocessor Test and Verification (MTV)* (2013)
25. U. Guin, M. Tehranipoor, D. DiMase, M. Megrdichian, Counterfeit IC detection and challenges ahead. ACM/SIGDA E-NEWSLETTER **43**(3) (March 2013)
26. U. Guin, M. Tehranipoor, On Selection of Counterfeit IC Detection Methods, in *IEEE North Atlantic Test Workshop (NATW)* (May 2013)
27. B. Hughitt, Counterfeit electronic parts, in *NEPP Electronics Technology Workshop* (June 2010)
28. U.S. Environmental Protection Agency, Electronic waste management in the united states through 2009 (May 2011)
29. BusinessWeek article and video from October 13, 2008, http://images.businessweek.com/ss/08/10/1002_counterfeit_narrated/index.htm
30. SAE, Counterfeit electronic parts; avoidance, detection, mitigation, and disposition (2009), http://standards.sae.org/as5553/
31. SAE, Test methods standard; counterfeit electronic parts. Work In Progress, http://standards.sae.org/wip/as6171/
32. CTI, Certification for coutnerfeit components avoidance program (September 2011)
33. IDEA, Acceptability of electronic components distributed in the open market, http://www.idofea.org/products/118-idea-std-1010b
34. Departnent of Defense, Performance specification: hybrid microcircuits, general specification for (2009), http://www.dscc.dla.mil/Downloads/MilSpec/Docs/MIL-PRF-38534/prf38534.pdf
35. J. Rhea, BAE Systems moves into third generation rad-hard processors (May 2002)
36. The Committee's Investigation into Counterfeit Electronic Parts in the Department of Defense Supply Chain. Senate Hearing 112–340 (November 2011), http://www.gpo.gov/fdsys/pkg/CHRG-112shrg72702/html/CHRG-112shrg72702.htm
37. CTI, Counterfeit Examples: Electronic Components (2013), http://www.cti-us.com/pdf/CCAP-101InspectExamplesA6.pdf
38. C. Mouli, W. Carriker, Future Fab: How software is helping Intel go nano–and beyond. IEEE Spectr. **44**(3), 38–43 (2007)
39. L.-T. Wang, C.-W. Wu, X. Wen, *VLSI Test Principles and Architectures: Design for Testability (Systems on Silicon)* (Morgan Kaufmann, San Francisco, 2006).

40. R.J. Abella, J.M. Daschbach, R.J. McNichols, Reverse engineering industrial applications. Comput. Ind. Eng. **26**(2), 381–385 (1994), http://dx.doi.org/10.1016/0360-8352(94)90071-X
41. M. Tehranipoor, F. Koushanfar, A survey of hardware trojan taxonomy and detection. IEEE Des. Test Comput. **27**(1), 10–25 (2010)
42. S. Adee, The hunt for the kill switch (May 2008), http://spectrum.ieee.org/semiconductors/design/the-hunt-for-the-kill-switch
43. CHASE, CHASE workshop on secure/trustworthy systems and supply chain assurance (April 2014), https://www.chase.uconn.edu/chase-workshop-2014.php
44. Defense Science Board (DSB), Study on high performance microchip supply (2005), http://www.acq.osd.mil/dsb/reports/ADA435563.pdf
45. Semiconductor Industry Association (SIA), Winning the battle against counterfeit semiconductor products (August 2013)
46. U.S. Department Of Commerce, Defense industrial base assessment: counterfeit electronics (January 2010)
47. H. Levin, Electronic waste (e-waste) recycling and disposal—facts, statistics & solutions (2011), http://www.moneycrashers.com/electronic-e-waste-recycling-disposal-facts/
48. L.W. Kessler, T. Sharpe, Faked parts detection. Print. Circuit Des. Fabr. **27**(6), 64 (2010)
49. J. Villasenor, M. Tehranipoor, Chop shop electronics. Spectr. IEEE **50**(10), 41–45 (2013)
50. SAE, SAE International, http://www.sae.org/
51. SAE, Fraudulent/counterfeit electronic parts; tool for risk assessment of distributors (2011), http://standards.sae.org/arp6178/
52. SAE, Fraudulent/counterfeit electronic parts: avoidance, detection, mitigation, and disposition - distributors counterfeit electronic parts; avoidance protocol, distributors (2012), http://standards.sae.org/as6081/
53. SAE, Fraudulent/counterfeit electronic parts: avoidance, detection, mitigation, and disposition—authorized/franchised distribution. Work In Progress, http://standards.sae.org/wip/as6496/
54. CTI, Components Technology Institute, Inc. http://www.cti-us.com/
55. B. Gassend, D. Clarke, M. van Dijk, S. Devadas, Silicon physical random functions, in *Proc. of the 9th ACM conference on Computer and Communications Security*. CCS '02 (ACM, New York, 2002), pp. 148–160
56. N. Robson, J. Safran, C. Kothandaraman, A. Cestero, X. Chen, R. Rajeevakumar, A. Leslie, D. Moy, T. Kirihata, S. Iyer, Electrically programmable fuse (efuse): from memory redundancy to autonomic chips, in *CICC* (2007), pp. 799–804
57. M. Miller, J. Meraglia, J. Hayward, Traceability in the age of globalization: a proposal for a marking protocol to assure authenticity of electronic parts, in *SAE Aerospace Electronics and Avionics Systems Conference* (October 2012)
58. C. Kuemin, L. Nowack, L. Bozano, N.D. Spencer, H. Wolf, Oriented assembly of gold nanorods on the single-particle level. Adv. Funct. Mater. **22**(4), 702–708 (2012)

Chapter 3
Counterfeit Defects

A wide variety of test methods are currently available for detecting counterfeit parts. The goal of these methods is to identify "defects" present in a part or a batch of parts under investigation. Counterfeit defects are those anomalies and changes that are not typically found in authentic parts. A counterfeit part may often contain one or more different anomalies and deviations from normal/usual form and/or functionality of a genuine component. These anomalies may be physical (i.e., related to the leads, package, etc.) or electrical (e.g., degradation in its performance or a change in its specifications). Since we assume that the foundries and assemblies have a fairly consistent manufacturing process and also comprehensively test their components, we should not expect any defects in genuine parts. Any anomalies or defective behaviors in a part must, therefore, be attributed to the part being counterfeit. The precise defects identified in a part can be linked to (i) the type of counterfeit. For example, recycled components undergo a harmful harvesting process that creates defects that are not likely present in overproduced, cloned, etc. counterfeit types; (ii) the expertise/capabilities of the counterfeiter. For example, some counterfeiters may have more expensive lasers to remark ICs and therefore may do so with greater precision that's more difficult to detect.

In this chapter, we will present all the defects and anomalies present in counterfeit components. We have classified them into four categories: procedural, mechanical, environmental, and electrical. It is important to understand these defects and anomalies before proceeding to later chapters which shall discuss the test methods (see Chaps. 4 and 5), assess the effectiveness of each test method (see Chap. 6 for the details), and discuss new advanced tests (see Chaps. 7 and 8).

M. Tehranipoor et al., *Counterfeit Integrated Circuits*,
DOI 10.1007/978-3-319-11824-6_3

3.1 Taxonomy of Counterfeit Defects

Counterfeit defects are those anomalies and changes that are not typically found in authentic parts. Anomalies vary based on size, shape, type, number, etc., depending on the capabilities possessed by the counterfeiters. The detection of one or more anomalies may be an indication that a component is counterfeit.

A taxonomy of the counterfeit defects was introduced in [1–3]. In this chapter, we present a revised and more comprehensive taxonomy of defects, which is divided into four categories: procedural, mechanical, environmental, and electrical as presented in Fig. 3.1. Each defect category is discussed in more detail in the sections below.

3.2 Procedural Defects

Procedural defects are related to the packaging and shipping of components and the markings of the component itself. The components should travel in the supply chain with proper protection against shipping, handling, and environmental conditions. Any damage due to the lack of protection may cause the components to fail during operation, and thus they must not be accepted as reliable components. A customer should receive documents verifying the authenticity of the components they purchased based on purchasing requirements. If there is a mismatch between the documents received compared to the original ones, then the batch of components would be flagged for further testing. Figure 3.2 shows the taxonomy of such defects. Each procedural defect is briefly discussed in the subsections below.

Invalid Lot Code (PP1)

A lot code is an identification number assigned to a lot of components manufactured by OCMs during a specified time period. It is typically placed on the outside of packaging. Using lot codes and tracking of components in the component supply chain are essential parts of good manufacturing practice. Any mismatch between the lot codes of a batch of parts under authentication and the lot codes stored in the OCM database should bring the batch of parts under suspicion for being counterfeit.

Invalid Packaging (PP2)

For this defect, the packaging does not satisfy the specifications of the OCM packaging. The OCM generally ships their product according to their standard shipping formats. For example, Intel ships their components using JEDEC standard specified trays or career tape/reel, and then places them in a conductive carbon

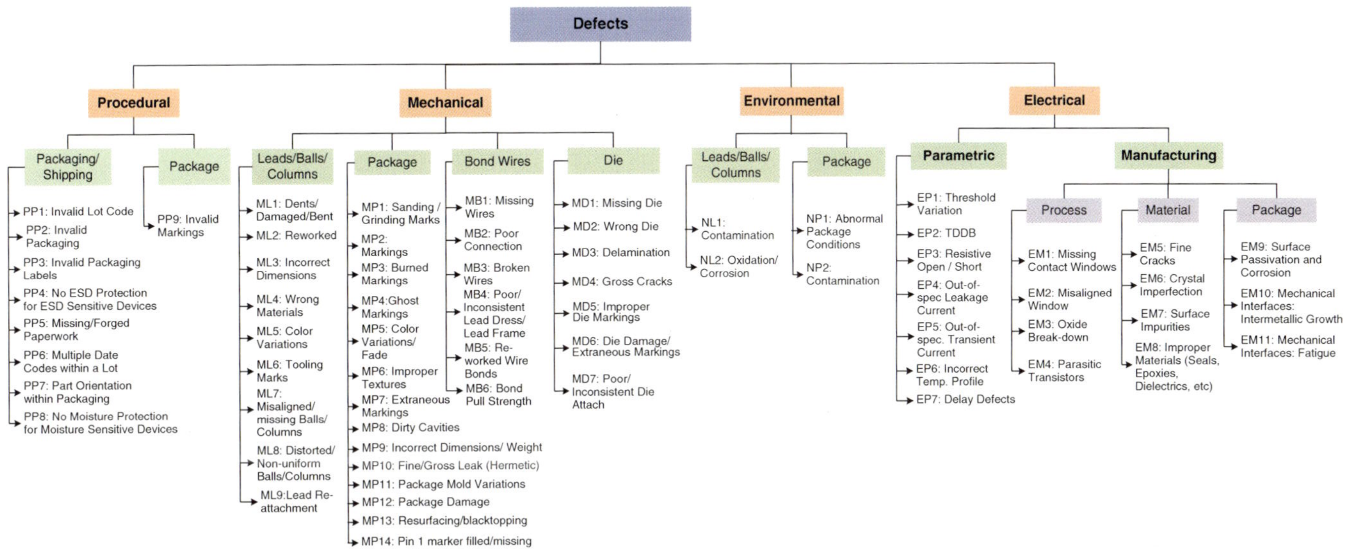

Fig. 3.1 A taxonomy of the defects and anomalies present in counterfeit electronic components

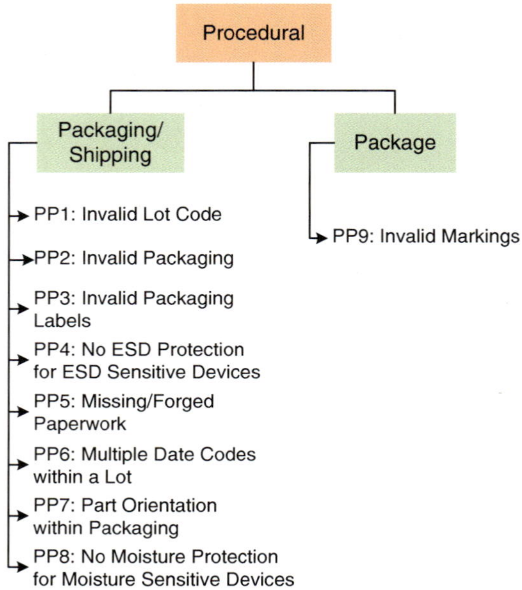

Fig. 3.2 Procedural defects

coated inner box [4]. The outer box provides protection during shipping, while the inner box protects the components from electrostatic discharge (ESD). If the components are found without proper packaging, the shipping, handling and/or environmental conditions may damage the components and cause them to fail during normal operation in the field. Hence, invalid packaging would not occur in the case of an authentic part.

Invalid Packaging Labels (PP3)

A mismatch between the shipping labels received compared to the OCM label signals a procedural defect. The package label for shipping is standardized nationally and internationally. The label exists in the form of product certifications, trademarks, proof of purchase, etc. Consumer use and safety information may also be required for display. Bar codes, universal product codes (UPC), and radio-frequency identification (RFID) labels are some common types of labeling. Figure 3.3 shows a packaging label displaying a grammatical error on it.

No ESD Protection for ESD Sensitive Components (PP4)

This defect exists when an electrostatic discharge (ESD) sensitive device ships without ESD protection. The inner layer of packaging boxes that contains these ESD sensitive components must be conductive, static dissipative, or antistatic in

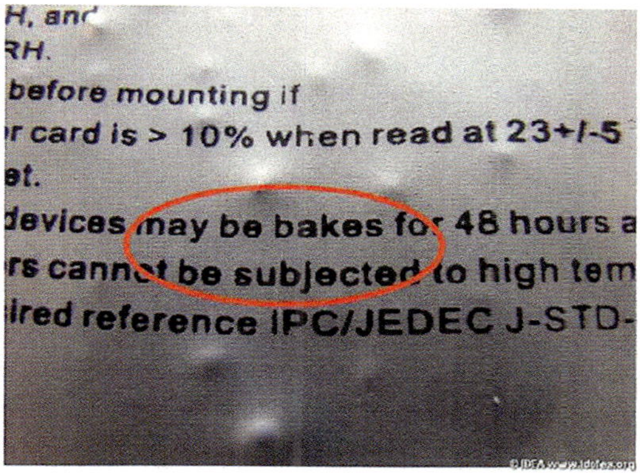

Fig. 3.3 Invalid packaging label [5]

order to meet EIA standard 541's electrostatic discharge requirements. A careless counterfeiter may ship an ESD sensitive counterfeit component without proper ESD protection.

Missing/Forged Paperwork (PP5)

Parts are generally shipped with test results and a part data sheet. The results can be made available for review from the OCM. This information could be missing altogether or forged by a counterfeiter.

Multiple Date Codes within a Lot (PP6)

The parts with same date code are generally packaged in a box shipped to the customer or distributor. Having different date codes on different components in the same lot raises the suspicion that they might have been collected from different sources and packaged together (such as for the recycled and remarked counterfeit types discussed in Chap. 2). Multiple date codes within a lot may provide an indication that the entire lot may be counterfeit.

Part Orientation Within Packaging (PP7)

The orientation of components in a packaging tape or reel needs to be checked. In an authentic lot, the parts are oriented in a similar fashion. Different part orientation in a package may lead to the suspicion that some components might have been replaced with counterfeit.

No Moisture Protection for Moisture Sensitive Devices (PP8)

This defect occurs when a moisture sensitive device (MSD) is shipped without moisture protection. The OCMs maintain their packaging formats to protect moisture sensitive devices. For example, Intel ships their moisture sensitive components within moisture barrier bags (MBBs) [4]. Each MBB contains: (i) one or more desiccant to absorb moisture captured inside it and (ii) humidity indicator card to display relative humidity level. A careless counterfeiter may ship an moisture sensitive counterfeit component without proper moisture protection.

Invalid Markings (PP9)

Markings on any component provide a detailed identity of the components. A detailed description of marking specifications can be found in Section 3.9.5 of MIL-PRF-38534H [6]. According to the description, "*Marking shall be in accordance with the requirements of this specification and the device specification. The marking shall be legible and complete, and shall meet the resistance to solvents requirements of method 2015 of MIL-STD-883*". The markings (see Fig. 3.4 for an example) of a component should contain the following unless otherwise specified, as described in [6]:

(i) Part or identifying number (PIN).
(ii) Index point.
(iii) Lot identification code or date code.
(iv) Device manufacturer's identification.
(v) Device manufacturer's CAGE CODE. The CAGE CODE online database is available at https://www.bpn.gov/bincs/begin_search.asp.
(vi) Country of manufacture.
(vii) Serialization, when applicable.
(viii) Special marking.
(ix) Electrostatic discharge sensitivity identifier.
(x) Certification mark.

Invalid markings on the package are clear evidence of counterfeiting. For example, invalid lot, date, or country code belongs to this category. Old parts are sometimes remarked with a current date code to make them look like currently manufactured parts (see Chap. 2 for more discussion on remarked parts). Figure 3.5 shows a part that was manufactured in the 47th week (November) of 2003. However, this part was received on June 3, 2003. This indicates that the part was supposedly manufactured 5 months after the date of receipt.

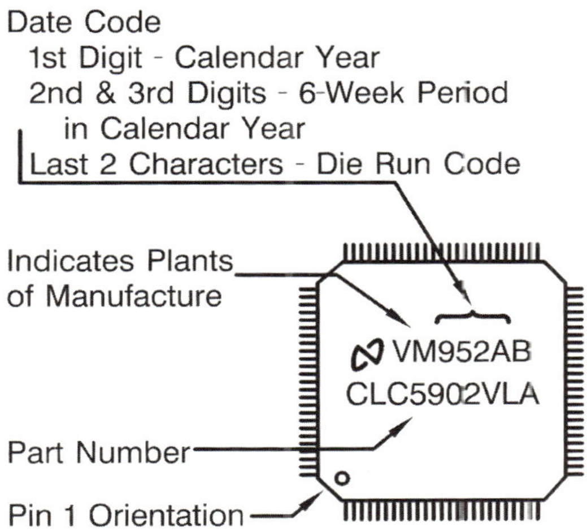

Fig. 3.4 Marking convention for a National Semiconductor's IC [7]

Fig. 3.5 Invalid date code [8]

3.3 Mechanical Defects

Figure 3.6 provides a detailed taxonomy of mechanical defects. Mechanical defects are directly related to a component's physical properties. Mechanical defects are categorized into four types—leads/balls/columns, package, bond wires and die—depending on the location of the defects. In the following, we will describe each of the defects listed in Fig. 3.6.

3.3.1 *Leads, Balls and Columns*

Leads, Balls or Columns on an IC can show how the part has been handled if it was previously used. Physically, leads should adhere to datasheet specifications such as, straightness, pitch, and separation. The leads' final coating should be consistent throughout the entire lot as well. Leads should also have a consistent elemental construction.

3.3.1.1 Dents/Damage/Bent (ML1)

Dents are unexpected impressions made in the surfaces of leads, balls, or columns by improper handling. For example, the regular shape of a lead can be bent or changed during the recycling process. This category includes all lead damage issues such as scratches, bent leads, broken leads, and missing leads. Figure 3.7 shows damaged leads of various parts.

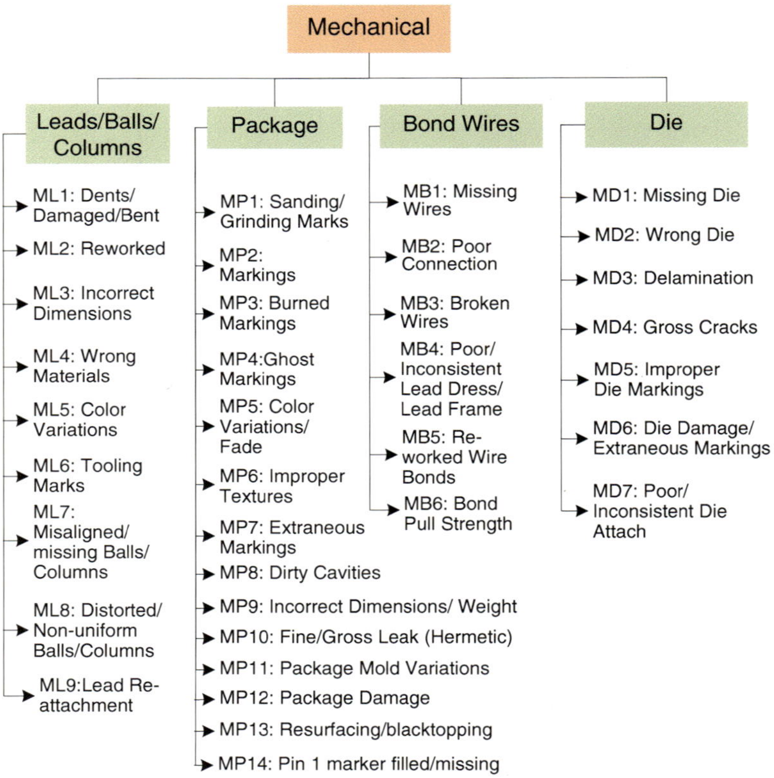

Fig. 3.6 Mechanical defects

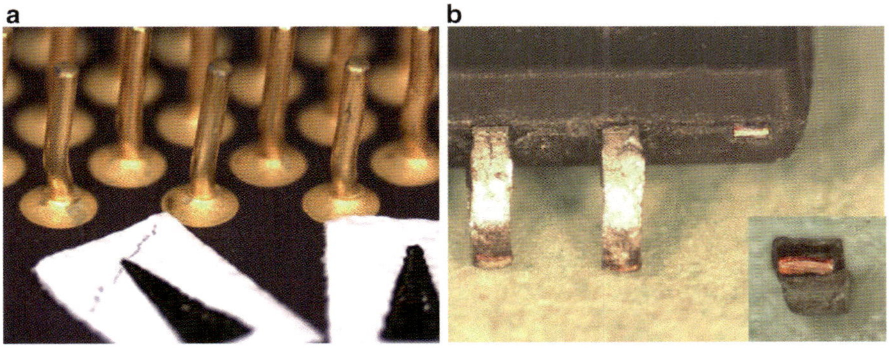

Fig. 3.7 Lead damage [*Source*: Honeywell Inc.] (**a**) Damaged columns (**b**) Broken leads

3.3.1.2 Reworked (ML2)

The reworking of leads defines this category of defects. Residual material on the lead, shown in Fig. 3.8a, b, clearly indicates the possibility of rework or reflow soldering having been performed on the leads. Replating of leads using tin (see Fig. 3.8c) also belongs to this category. Reworking may be performed on a recycled or remarked chip.

3.3.1.3 Incorrect Dimensions (ML3)

Physically, leads, balls, and columns should adhere to datasheet specifications, including straightness, pitch, separation, etc. Any mismatch in these specifications would lead to this type of defect. Counterfeit chips that belong to recycled, remarked, cloned, or overproduced types (see Chap. 2) may have incorrect dimensions if the counterfeiters do not adhere to the datasheet specifications

3.3.1.4 Wrong Materials (ML4)

Whenever the chemical composition of leads, balls, and columns differs from the specification sheet, the case would be placed in this category of defect. For example, if a lead's plating was supposed to be nickel and turned out to be tin, then it would be apparent that the wrong material had been used. The hazardous element lead (Pb) is not supposed to be present on the leads of an IC. Figure 3.9a shows that Pb is detected (encircled in black) during energy dispersive spectroscopy (EDS), a test used to detect lead finish.

Wrong materials could appear in various counterfeit types discussed in Chap. 2: (i) leads of a recycled chip can be reworked if corroded or otherwise damaged, (ii) a remarked chip may only have its markings changed while the leads might not have

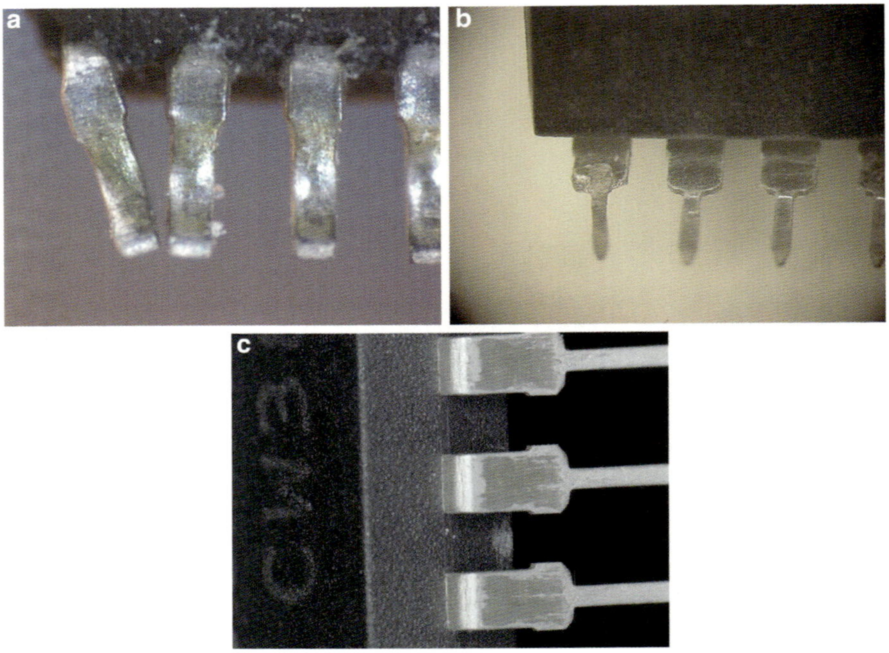

Fig. 3.8 Reworked leads. (**a**) Residual flux on the leads (*Source*: Honeywell Inc.), (**b**) repaired and reflowed leads (*Source*: Honeywell Inc.), (**c**) poorly re-tinned

been reworked to match the spec sheet, (iii) a cloned or overproduced chip may be created without adhering to the datasheet specifications.

3.3.1.5 Color Variations (ML5)

If the leads' color differs from the datasheet specifications (or from authentic parts), then this could signal that the lead has been reworked. If the leads appear to have a darker or duller finish, it could be a sign that they have been soldered or removed from a previously used printed circuit board (PCB). Figure 3.10 displays this defect.

3.3.1.6 Tooling Marks (ML6)

Missing tool marks on the lead may indicate the presence of a used component, as replating often covers tool marks during recycling. Insertion marks on leads may suggest that the part was previously installed elsewhere. The leads of an authentic part may contain tooling marks. The presence (the authentic part does not contain

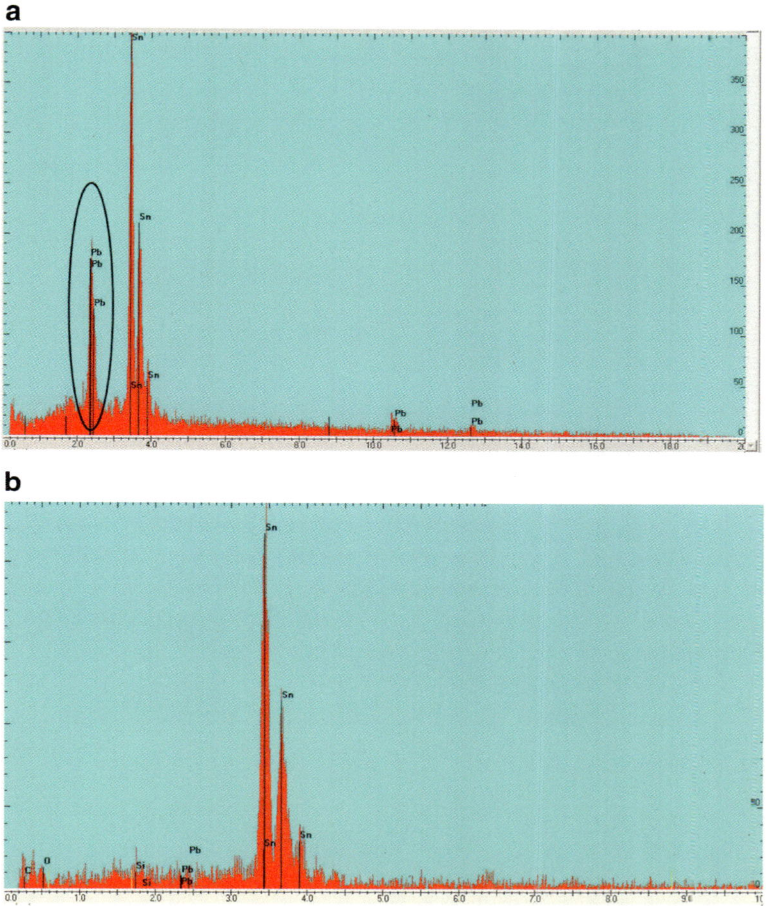

Fig. 3.9 Wrong materials. (**a**) Counterfeit: lead (Pb) found, (**b**) genuine: no lead (Pb) found

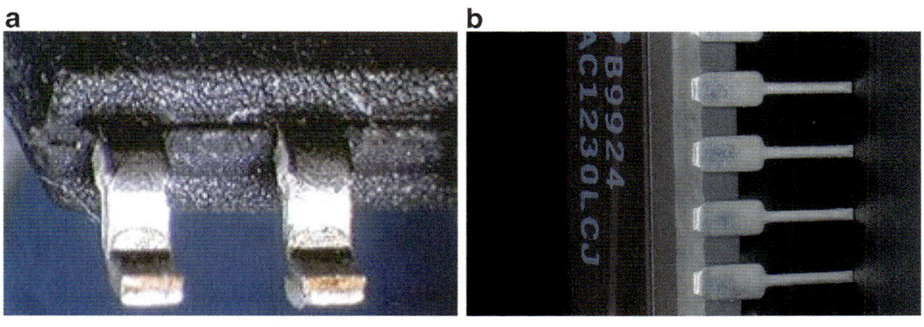

Fig. 3.10 Color variations (**a**) Residual solder material on leads [9] (**b**) Color variations on leads

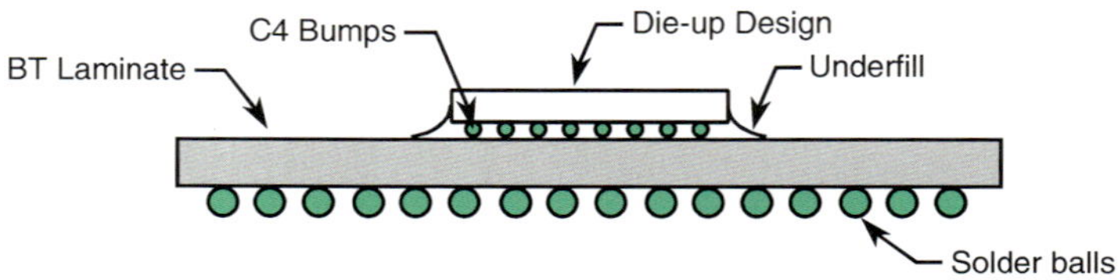

Fig. 3.11 Cross sectional view of a BGA packaged IC [10]

Fig. 3.12 Missing solder balls [5]

tooling marks) or absence (the authentic part contains tooling marks) of these marks could be an indicator of counterfeiting.

3.3.1.7 Misaligned/Missing Balls/Columns (ML7)

Array packaging technology has become popular among the semiconductor industry today as they provide a huge increase in input/outputs (I/O) pin density for large integrated circuits. Ball grid array (BGA) and column grid array (CGA) are two popular surface-mount packaging techniques used for ICs. These arrays provide not only higher interconnect densities, but also solve the problems associated with leads (such as bent, broken, etc.). Figure 3.11 shows a cross sectional view of a BGA packaged IC.

When the array-packaged components are recycled, the solder balls or columns may get misaligned or some of them may be missing (see Fig. 3.12). Misaligned or missing balls or columns are an indicator of potential reworking that was done during or after the recycling process.

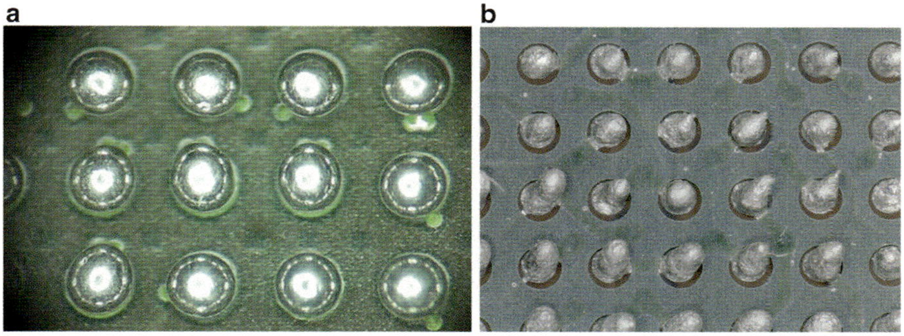

Fig. 3.13 Damage or altered BGA Balls. (**a**) Distorted solder balls (*Source*: Honeywell Inc.); (**b**) Non-uniform/damaged solder balls

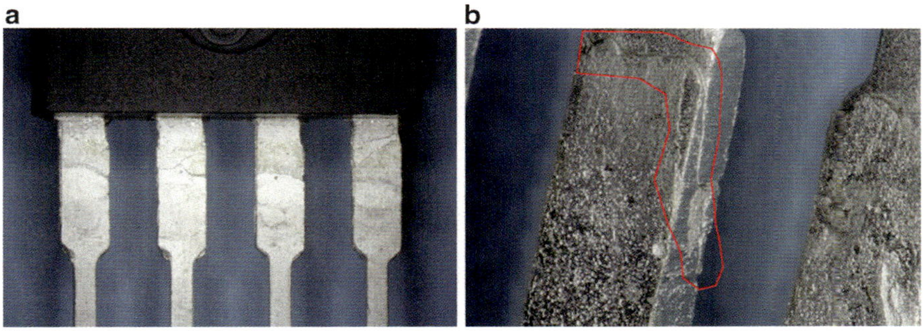

Fig. 3.14 Lead re-attachment [*Source*: Honeywell Inc.] (**a**) Reattached leads (**b**) Magnified joint location of one of the leads

3.3.1.8 Distorted/Non-uniform Balls/Columns (ML8)

Non-uniform or distorted balls and columns may indicate that the part is counterfeit. Solder balls should not show signs of reworking. Figure 3.13a shows teardrop shaped solder balls. Figure 3.13b shows damaged solder balls where the residual solder materials are clearly visible.

3.3.1.9 Lead Re-attachment (ML9)

Lead re-attachment is the process by which the original leads or ball grid array (BGA) are removed and replaced with new or different leads/BGA to make a part lead-free. It also involves the reattachment of a lead that was damaged or broken during recycling process. Some possible indicators of this kind of counterfeiting would be evidence of visible oxidation; differences in leads, balls and columns' size and texture; and leads, balls and columns not passing solderability testing. Figure 3.14 shows that new leads have been attached to the old broken leads. The solder material is clearly visible between the joint location (see Fig. 3.14b).

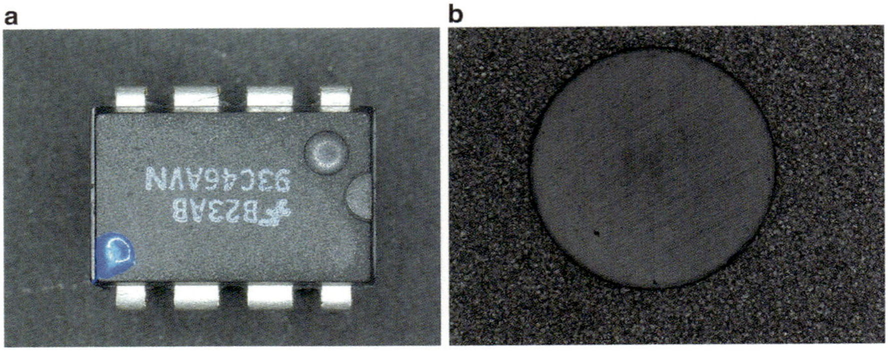

Fig. 3.15 Sanding marks. (**a**) Sanded top surface, (**b**) sanded dimple

3.3.2 *Package*

The package of an IC provides significant information regarding its authenticity. For example, country of origin, date and lot codes, device manufacturers' identification, etc. (see Defect 3.2) are marked on the package. If the package exhibits any sanding or grinding marks externally, it has likely been recycled or remarked. Further inspection for the blacktop coating should be done to determine whether this is the case. Ghost markings, color variations, improper textures, and extraneous markings on the package clearly indicate that a part has been reused.

There are several package type present today depending on epoxy-molding compound (EMC), ceramic, metal, or engineering thermoplastic materials [11]. A complete chip encapsulation is performed by whereas the other materials create open-cavity packages. It is reported in [11] that epoxy based packaging accounts for more than 90 % of all semiconductor packaging as it is the least expensive among all. On the other hand, full hermetic packages are typically made of ceramic, metal or both [12], which provide protection against not only gasses and moisture, but also shipping and handling. In the following, we will describe all the package defects,

3.3.2.1 Sanding/Grinding Marks (MP1)

If the package exhibits any sanding or grinding marks externally, it has likely been remarked. Generally, counterfeiters use sand blasting processes to remove markings from the package. This process generally results in a distinct visual signature being present on the package. Figure 3.15 shows sanding marks on a counterfeit IC.

3.3.2.2 Markings (MP2)

Markings should be consistent and conform to the authentic parts. Marking defects occur when there is a mismatch between the marking of a part under authentication

Fig. 3.16 Marking defects (**a**) Encircled marking near Tundra logo (**b**) Encircled marking near mold mark

and the authentic part. The markings on the package should be permanent and clean with no indications of remarking.

Figure 3.16 shows two different counterfeit "TUNDRA PCI to Motorola Processor Bridge" components. These components are blacktopped and then remarked. The country of origin, shown in the top left corner of the component, is not placed in the exact same location. It is close to the TUNDRA logo in Fig. 3.16a, whereas it is closer to mold mark in Fig. 3.16b. The quality of the marking is also poor for these components, as we can observe that corners of the letters are not sharp and most of the letters contain holes when viewed under larger magnification.

3.3.2.3 Burned Markings (MP3)

Laser machining is one of the most popular technologies for package markings. The imprecise exposure of the laser beam may cause burn marks on the package during the remarking process (see Fig. 3.17).

3.3.2.4 Ghost Markings (MP4)

Ghost marking occurs when the counterfeiters do not entirely remove the original marking before printing the new one. The part markings are faintly visible and appear behind the new markings on the parts. The original part markings can be seen either under low-power magnification or after marking permanency testing. In Fig. 3.18a, the original white markings are clearly visible. Figure 3.18b shows a ghost marking, remnants of the original marking, on the right side of the component.

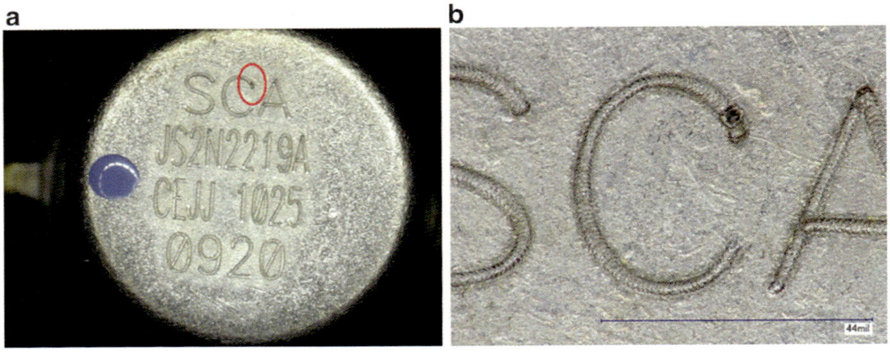

Fig. 3.17 Burned markings [*Source*: Honeywell Inc.] (**a**) Burned mark encircled in red (**b**) Magnified location of the burned mark

Fig. 3.18 Ghost marking (**a**) New marks on top of residual original marks (*Source*: Honeywell Inc.) (**b**) Faded marks encircled in red

3.3.2.5 Color Variations/Fade (MP5)

If the package color is faded, it would be a clear indication of part counterfeiting. Figure 3.19a shows the variations of color in the same lot of Intel processors. The heat sink witness mark is another example of this type of defect. Light blackish marks on the Intel processor show the prior attachment to a heat sink, which is shown in Fig. 3.19b.

3.3.2.6 Improper Textures (MP6)

The mismatch of textures between top, side, and bottom surfaces of a package belongs to this category. Black topping is performed to hide the sanding or grinding marks on the top surface. If a part is blacktopped, then there will be a clear texture

Fig. 3.19 Color variations [*Source*: Honeywell Inc.]. (**a**) Color variations, (**b**) heat sink mark

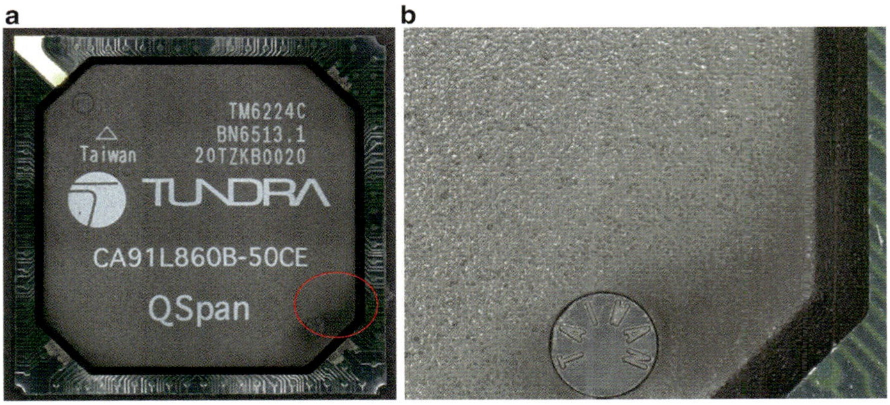

Fig. 3.20 Improper texture (**a**) Improper texture encircled in red (**b**) Magnified location of the corner

difference between the top and the side, as well as the top and bottom surfaces. Figure 3.20b, zoomed over Fig. 3.20a, shows a different surface texture (blackish and smooth) near the mold mark and the corner compared to the central part of the component

3.3.2.7 Extraneous Markings (MP7)

Surface scratches (see Fig. 3.21a) and unexpected ink dots are examples of this type of defects. This includes scratch marks under solder balls and columns on CGA/BGA parts (see Fig. 3.21b).

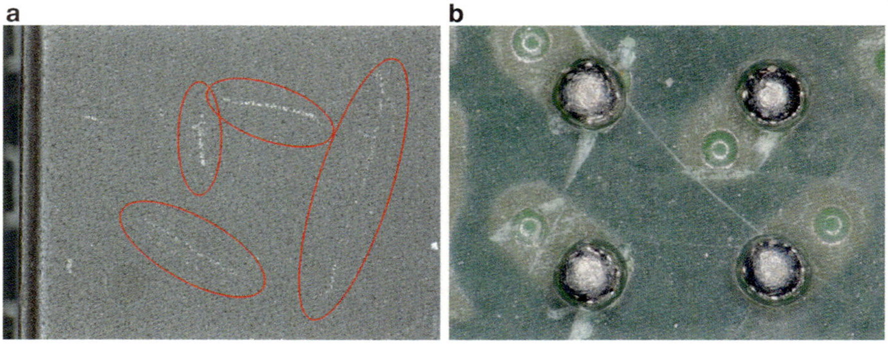

Fig. 3.21 Extraneous markings. (**a**) Scratches on the package, (**b**) scratches on the substrate of flip-chip BGA package (*Source*: Honeywell Inc.)

3.3.2.8 Dirty Cavities (MP8)

This category of defects consists of cavities in hermetically sealed components that are not clean and have extra materials in them. Cavities such as hermetic cavity, are commonly integrated into an IC component to offer protection against mechanical and handling stress, especially for components with fragile surface features.

3.3.2.9 Incorrect Dimensions/Weight (MP9)

Package dimensions should be consistent with and conform to the part data sheet. Due to resurfacing and blacktop coating, the weight of the part may vary. Figure 3.22 shows an uneven edge bevel due to sanding.

3.3.2.10 Fine/Gross Leak (Hermetic) (MP10)

The seal on hermetic parts is a crucial component that ensures a part's proper functioning within the environment for which it was designed. The seal of a hermetic part can be broken by excessive force or heat, both of which are typical of a crude recycling process.

3.3.2.11 Package Mold Variations (MP11)

Molding compounds are generally composite materials consisting of epoxy resins, phenolic hardeners, silica, catalysts, pigments, mold release agents, etc. Variations from the data sheet (authentic or genuine part) can confirm a part as counterfeit. Defects involving different package mold shapes within the same lot also belong

Fig. 3.22 Incorrect dimensions [*Source*: Honeywell Inc.]

a **b**

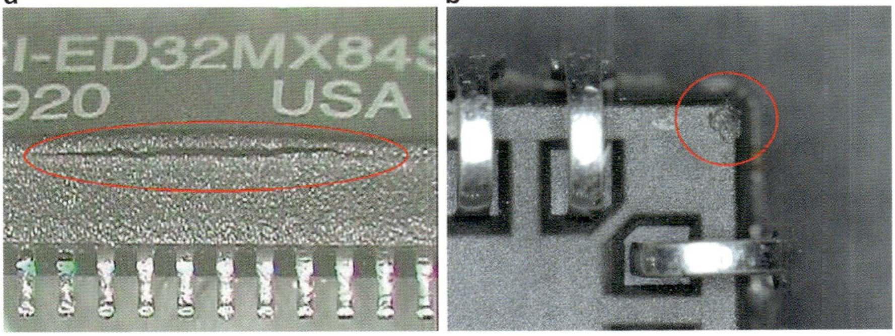

Fig. 3.23 Package damage. (**a**) External crack [13], (**b**) chipout at the corner

to this category. Recycled, remarked, overproduced, and cloned counterfeit types could potentially have this defect.

3.3.2.12 Package Damage (MP12)

If the package exhibits damages such as cracks or chipout, then it would fall within this category of defect. Figure 3.23a shows an external crack on a package. Figure 3.23b displays a corner chipout of an IC due to improper handling.

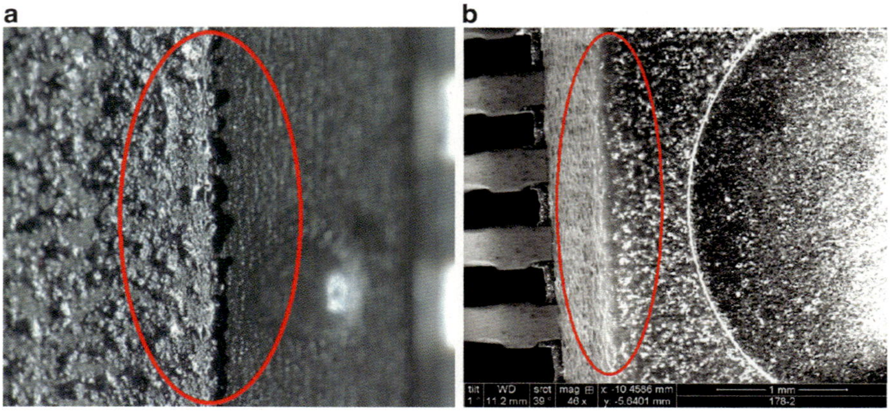

Fig. 3.24 Resurfacing/blacktopping (**a**) Optical image (**b**) SEM image

3.3.2.13 Resurfacing/Blacktopping (MP13)

If the package has a secondary coating applied to one or more sides of the package, then it has probably been resurfaced or blacktopped. The coating is intended to obscure the original package features and provide visual consistency to the lot. Figure 3.24 shows the exposed sanding marks after removing blacktop coating.

3.3.2.14 Pin 1 Marker Filled/Missing (MP14)

Pin 1 marker shows the location of the pin 1 of a component. Pin 1 marker or mold markers (dimples) on a package may: (i) have a production number or letter, (ii) be missing, (iii) be filled by "black-topping", or (iv) not be shiny. Figure 3.25 shows that the dimples are very shallow and have sanding marks.

3.3.3 Bond Wires

An integrated circuit contains a die, bond wires, and a structure that is used to hold them all in place, which is shown in Fig. 3.26. These internal pieces could potentially provide significant information regarding a component's authenticity. It is important to observe the shape, size and count of the bond wires. It is common for some circuits to bi-wire the power and ground connections for better current carrying capability. The absence of any of these bond wires, or broken bond wires will definitely cause a failure of a component in the field. The defects related to bond wires are listed below.

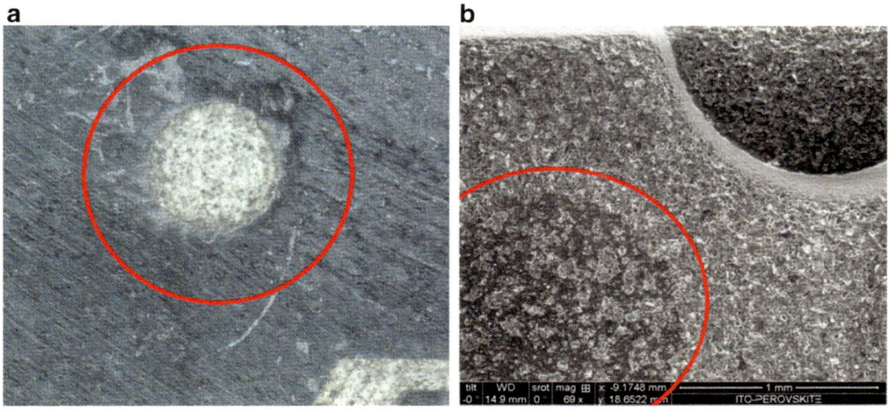

Fig. 3.25 Sanded and partially filled dimples. (**a**) Sanded dimple, (**b**) filled dimple

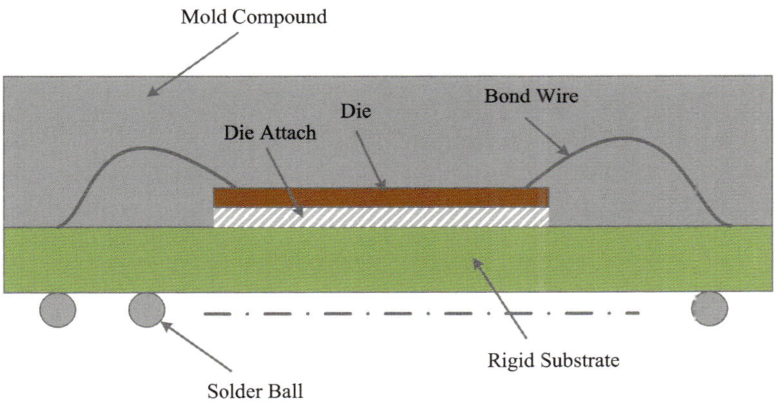

Fig. 3.26 A wire-bond packaged chip

3.3.3.1 Missing Wires (MB1)

The inside of an integrated circuit (IC) contains a die and bond wires. If some of the bond wires are missing, the circuit will fail during functional operation. Usually, multiple bond wires are used for a single connection to support the current delivered to the die. The component will work as specified, but, it may fail under extreme conditions. This might be an indication of a recycled component, where the die is completely repackaged or a cloned component, where the counterfeiter use single wire bond. Figure 3.27a illustrates that there are no bond wires inside the package, whereas Fig. 3.27b shows its authentic counterpart.

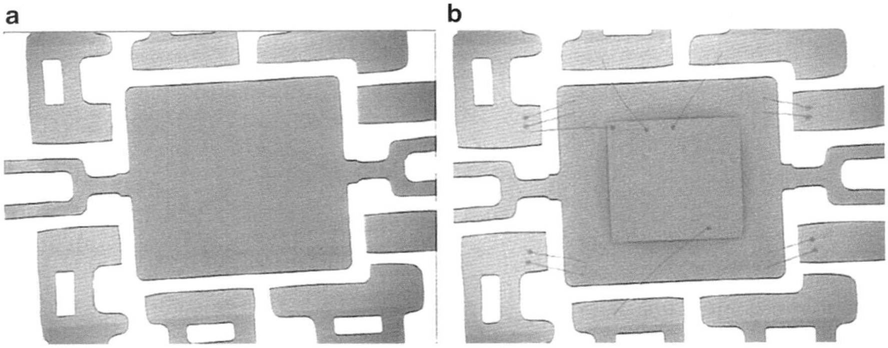

Fig. 3.27 Missing bond wires [8]. (**a**) Counterfeit, (**b**) authentic

3.3.3.2 Poor Connection (MB2)

A defect occurs when the connection between the die and leads, balls or columns of a component is poor. This is a type of latent defect, where the part may work normally for a while before the user experiences degradation in performance. If under enough environmental stress or if exposed to a large shock (ESD, for example), the wire itself may completely burn out. This defect may occur in recycled components where the connections are degraded due to the prior usage. This may also be visible in out-of-spec/defective components. Cloned components may also possess this defect when the counterfeiter uses inferior materials for cloning.

3.3.3.3 Broken Wires (MB3)

Components that have gone through the recycling process (see Chap. 2 in detail) may have been mishandled to the extent that the connection from the bond wire to the die is broken. Figure 3.28 illustrates several broken bond wires inside the package. This defect may also occur in out-of-spec/defective components when it is escaped from assembly after packaging process.

3.3.3.4 Poor/Inconsistent Lead Dress/Lead Frame (MB4)

The presence of inconsistent lead dress serves as a clear indication that a part is counterfeit. The presence of two different types of lead frame structures for chips that are supposed to belong to the same lot is also a good indicator of parts being counterfeit. Figure 3.29 shows two different lead frame structures for two counterfeit Intel TB28F400B5-B80 flash memories. The die in Fig. 3.29b is rotated 180° compared to the die in Fig. 3.29a.

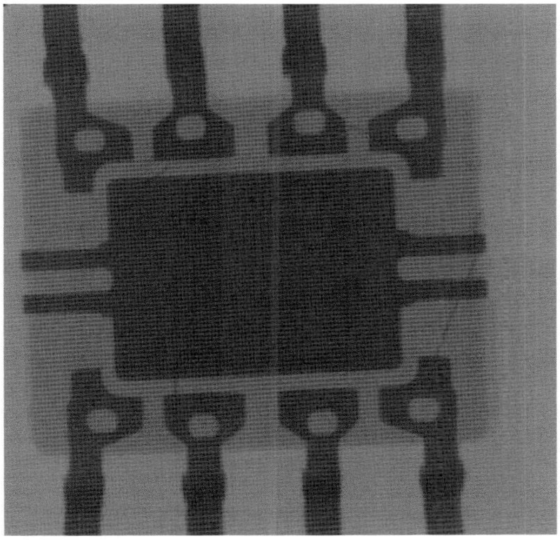

Fig. 3.28 Broken bond wires [*Source*: Honeywell Inc.]

3.3.3.5 Re-worked Wire Bonds (MB5)

The die is lifted from the package during die recovery and then re-packaged to a new component. The counterfeiters re-ball the bond wire terminations leaving double ball bonds (see Fig. 3.30). In an authentic component, one would normally see single ball bonds. Double ball bonds are allowed up to certain amounts for re-work, but when all of them are double bonded the part is likely counterfeited by die recovery.

3.3.3.6 Bond Pull Strength (MB6)

The strength of the bond wire interconnection system, including the strength of the bond wire and the strength of the bond wire interconnection to the pad, can help gauge whether or not counterfeiting has taken place. This defect can occur in a counterfeit (e.g., recycled, out-of-spec/defective, cloned, and overproduced) part having improper bond wires or an intermetallic growth formation at the connection pad, which causes a poor connection. In the worst case, the bond wires may be detached from the die (see Fig. 3.31).

3.3.4 Die

Die is a semiconducting material on which an electronic circuit is fabricated. It is extremely important to inspect the die very carefully. In one shocking case, a component was shipped with no die inside the package [8]. Admittedly, this

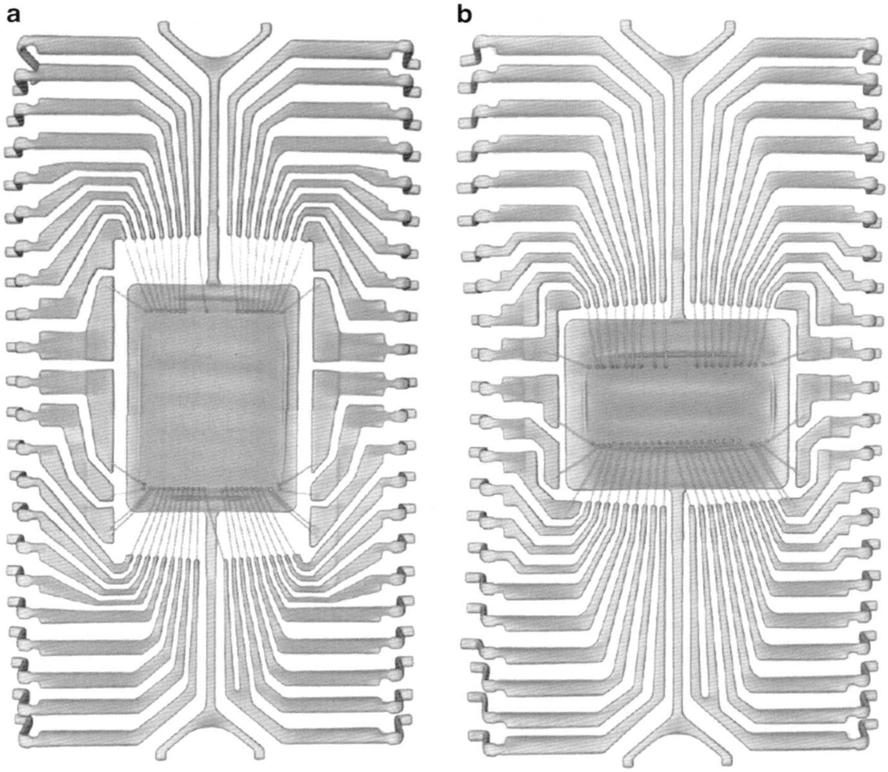

Fig. 3.29 Different lead frame structures (**a**) Authentic die (**b**) Rotated die

scenario does not happen often. A more possible scenario is the likelihood of a component having a different die than the one indicated by the markings on the package. This could be the case if the package has been remarked or if a different die was transplanted into a new package. There are markings on the die that can help in proving authenticity when they are compared to the package markings. These defects are described below:

3.3.4.1 Missing Die (MD1)

This defect occurs when the die is missing inside the package. Figure 3.27a showed the X-ray image of a counterfeit IC where no die inside the package is observed.

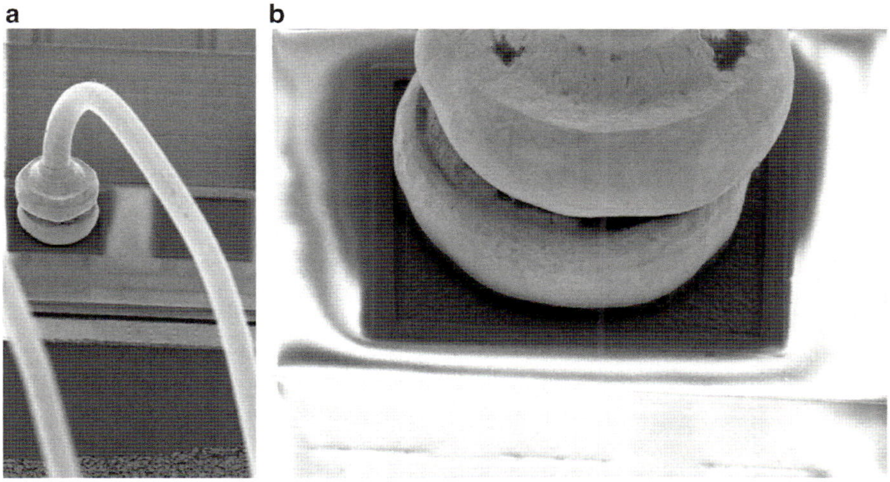

Fig. 3.30 Double ball bonds [13] (**a**) Bond wire having double ball bonds (**b**) Magnified location of the joint

Fig. 3.31 Lifted bond wires [13]

3.3.4.2 Wrong Die (MD2)

In this case, the die is different from what it is expected to be. Different die orientations existing within a single lot would also come under this category. Figure 3.32 shows an Intel die found in an AMD package.

Different die sizes within the same lot may be an indication of a wrong die. However, the die sizes may vary within different lots, as the OCM may adopt a newer process (for example, the die size will be smaller when manufactured with the newer 45 nm technology node compared to older 90 nm technology node). Similarly, different die layouts within the same lot may also imply a wrong die. However,

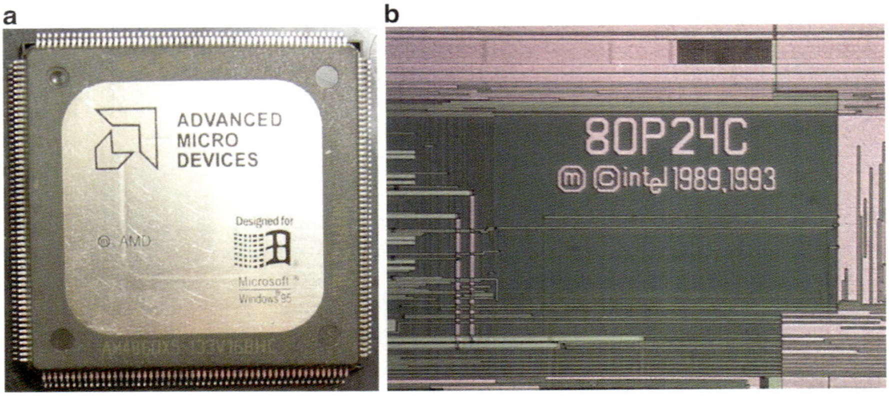

Fig. 3.32 Intel die inside an AMD package [13]. (**a**) AMD package, (**b**) Intel die

different die layouts for different lots may not necessarily suggest a wrong die as the OCM may have several layout versions of the same design.

3.3.4.3 Delamination (MD3)

Due to imperfections in fabrication, a die may contain trace amounts of air between its layers. When heated, these pockets will expand. If there is enough air present, the die pocket will expand to the point of delaminating, a process by which adjacent, connected layers of the die separate causing the circuit to open in that area. This is known as "popcorning" due to the resemblance.

3.3.4.4 Gross Cracks (MD4)

A component that has gone through a crude recycling process is subjected to extreme changes in temperature and harsh environments that it was not designed to withstand. If gross cracks exist in the die, then the defects come under this category.

3.3.4.5 Improper Die Markings (MD5)

There are markings on the die that can help in proving authenticity when compared to the package markings. When the marking on the die does not match the marking on the package, there may be a high possibility of that part being counterfeit. However, the part may be still be authentic if the OCM uses different markings for the die and the package.

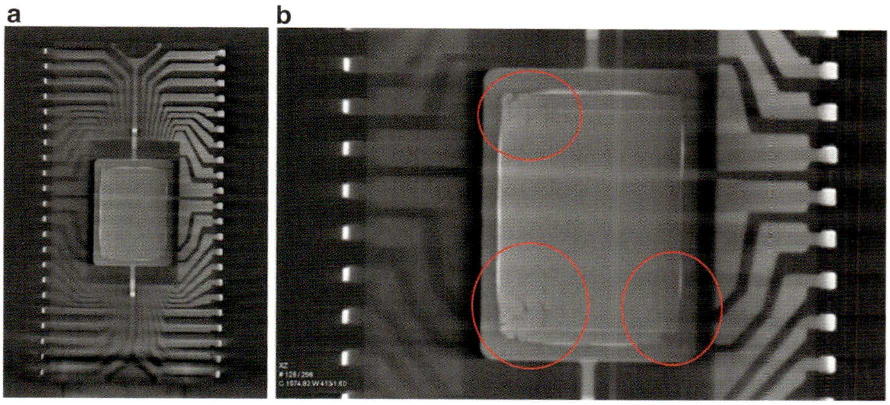

Fig. 3.33 Damaged die (a) X-Ray image of an Intel flash memory (b) Damaged corner encircled in red

3.3.4.6 Die Damage/Extraneous Markings (MD6)

The die may be damaged during the recycling process. Extraneous marks, such as scratches, may also be present. Figure 3.33 shows a damaged die (see Fig. 3.33b) inside an Intel TB28F400B5-B80 flash memory (see Fig. 3.33a). The voids in the X-ray images indicate the damages caused during harsh recycling process.

3.3.4.7 Poor/Inconsistent Die Attachment (MD7)

Excessive voids in die attachment or inconsistent die attachment could be a sign of reworked or cloned products.

3.4 Environmental Defects

Environmental defects are caused when the environmental parameters interact with the outer structure of a component. Oxidation and corrosion on leads are caused when a part is kept for a long time without proper protection. In addition, during the recycling process, the leads can easily get oxidized at higher temperatures and contaminated by other materials. Figure 3.34 shows the classification of environmental defects.

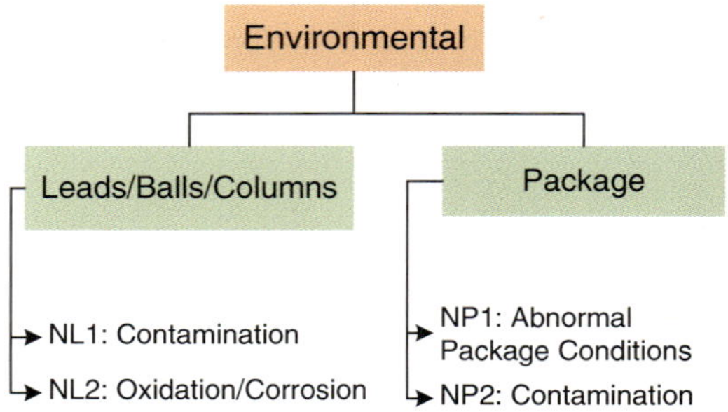

Fig. 3.34 Environmental defects

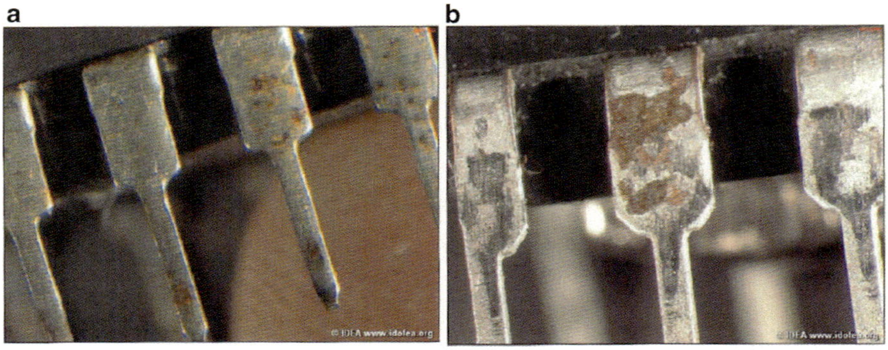

Fig. 3.35 Oxidation on the leads of an IC [5] (**a**) Lead contamination by oxidation (**b**) Oxidation on shoulder of lead

Contamination (NL1)

Leads, balls, or columns are contaminated by hazardous substances (Restriction Of Hazardous Substances, ROHS). If there is contamination, then the plating may be correct, but it may have organics over it.

Oxidation/Corrosion (NL2)

There may be oxidation, or corrosion on the lead due to the harsh recycling process. Whiskers could indicate abnormal storage or lead frame material. Figure 3.35 shows the oxidation, or corrosion on the leads, which is a clear indication of that the component was stored without proper protection against environmental conditions.

Fig. 3.36 Corrosion on the package of an IC [*Source*: Honeywell Inc.]

Abnormal Package Conditions (NP1)

Abnormal package conditions could include such items as presence of corrosion (see Fig. 3.36), which can cause significant damage to a package during the harsh recycling process.

Contamination (NP2)

The package of a component is made of epoxy-molding compound, ceramic, metal, or engineering thermoplastic materials [11]. When component are kept in open air it may contaminated. It may also be contaminated during the recycling process.

3.5 Electrical Defects

A defect in an electronic system is the unwanted difference between the implemented hardware and its intended design [14]. Typical electrical defects of the counterfeit components can be classified into two distinct categories. These are parametric defects and manufacturing defects, both of which are shown in Fig. 3.37. Functional or parametric failures occur because of electrical defects. Unlike the procedural, mechanical, and environmental defects explained above, it is hard to visualize most types of electrical defects. A detailed description of these defects can be found in [14, 15].

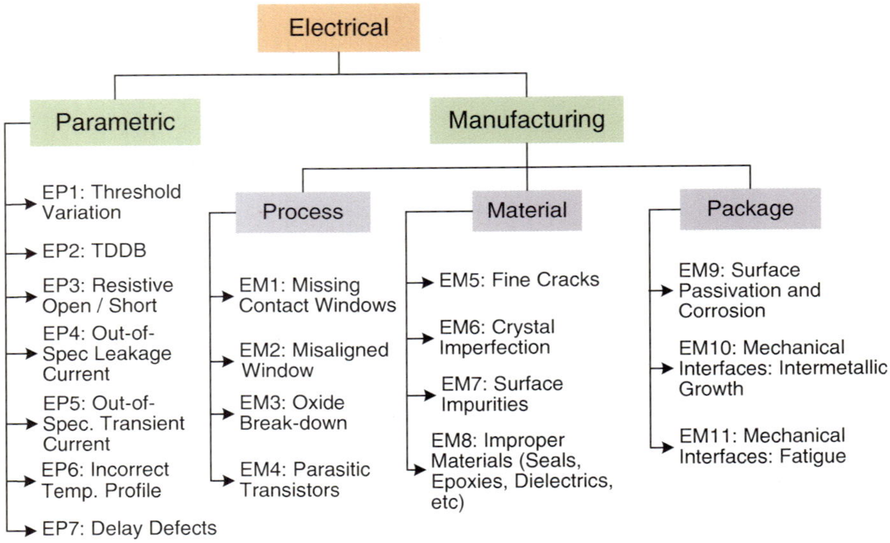

Fig. 3.37 A taxonomy of electrical defects

3.5.1 Parametric Defects

Parametric defects represent a shift of component parameters from what is expected. For example, a shift in circuit parameters due to aging will occur when a chip is used in the field for some time. Aging of a chip used in the field can be attributed to four distinct phenomena, which are becoming more prevalent as feature size shrinks. The most dominant phenomena is negative bias temperature instability (NBTI) [16–20] and hot carrier injection (HCI) [20–23] which is prominent in PMOS and NMOS devices, respectively. NBTI occurs in p-channel MOS devices stressed with negative gate voltages and elevated temperature due to the generation of interface traps at the Si/SiO_2 interface. Removal of the stress can anneal some of the interface traps, but not completely. As a result, it manifests as the increase in threshold voltage (V_{th}) and absolute off current (I_{off}) and the decrease in absolute drain current (I_{DSat}) and transconductance (g_m). HCI occurs in NMOS devices caused by the trapped interface charge at Si/SiO_2 surface near the drain end during switching. It results in non-recoverable V_{th} degradation. These effects also lead to out-of-spec leakage current and out-of-spec transient current. Delay defects are also the direct effect of all the parametric variations mentioned above.

Time-dependent dielectric breakdown (TDDB) [24, 25] is another effect of aging that irreparably damages MOS devices. MOS devices with very thin oxide layers are generally subjected to very high electric fields. The carrier injection with this high electric field leads to a gradual degradation of the oxide properties, which eventually results in the sudden destruction of the dielectric layer. Finally, electromigration

[26] the mass transport of metal film conductors stressed at high current densities—may cause the device to fail over time. If two interconnects are close enough, the migration of atoms can lead to bridging between them. Open circuits may also result due to the apparent loss of conductor metal. Depending on the circuit's workload (input combinations, temperature, environmental noise, etc.) and technology node, the amount of degradation a chip experiences will be different.

These parametric defects may potentially be present in all counterfeit types. For example, the device parameter shift due to the usage of a component shall lead to electrical defects present in recycled components. For overproduced and out-of-spec/defective components, these defects may be the manifestation of manufacturing variability. An untrusted entity may source defective components in the supply chain which possess opens, shorts, etc. Finally, electrical defects may also be present in cloned components, which may have been manufactured with different technology node or not tested properly.

In the subsections below, we will briefly discuss each parametric defect.

3.5.1.1 Threshold Variation (EP1)

This defect occurs when the input low voltage (V_{IL}) and input high voltage (V_{IH}) specified in the datasheet, do not make the device output change (high to low or low to high). These voltages generally have a lower value in the lower technology node. For example, V_{IL} and V_{IH} of a component manufactured with a 90 nm technology node will be higher compared to the same component manufactured with 45 nm technology nodes.

3.5.1.2 Time-Dependent Dielectric Breakdown (TDDB) (EP2)

TDDB occur when a dielectric (gate oxide) under a constant electric field (which is less than the dielectric breakdown strength) breaks down with time. The carrier injection with this high electric field leads to a gradual degradation of the oxide properties, which eventually results in the dielectric layer's sudden destruction.

3.5.1.3 Resistive Open/Short (EP3)

Resistive open or short may result from electromigration when the component is used in the field. Electromigration refers to the mass transport of metal film conductors that are stressed by high current densities and aging, causing a device to fail over time. Electromigration often occurs in aluminum (Al) wiring on components. If the two interconnects are close enough, the atoms may migrate such that the interconnects become bridged (i.e., creating a short circuit). During the current flow, Al wire degrades when the electrons collide with Al grains of the wire. The wire are eventually burnt out due the dislocation of Al grains. Components that are remarked to a higher grade or recycled may experience electromigration that leads to unexpected failure.

3.5.1.4 Out-of-Spec Leakage Current (EP4)

Leakage current refers to the undesired current that flows while a CMOS device is in an ON or OFF state. The leakage current of a counterfeit component may be different from those obtained from authentic parts or datasheet specifications. For example, components manufactured with different technology nodes may lead to this defect. Generally, the components manufactured with lower technology nodes have higher leakage current [27]. For example, the leakage current of a component fabricated with 45 nm process is higher compared to a component manufactured with 90 nm process. In addition, leakage current generally decreases due to the increase of threshold voltage (V_{th}) when the device ages in the field [28].

3.5.1.5 Out-of-Spec Transient Current (EP5)

The transient current is the current flowing in a CMOS circuit when it switches in between ON and OFF states or vice versa. If the transient current is different from those obtained from authentic parts or datasheet specifications, this is considered another defect. Components with different technology provide different transient current. For example, the transient current of a component manufactured with a 90 nm technology node will be different compared to the same component manufactured with any lower technology nodes. This defect may also arise from the manufacturing process of a component. This defect can potentially be present in all counterfeit types.

3.5.1.6 Incorrect Temperature Profile (EP6)

A circuit workload (input combinations, temperature, environmental noise, etc.) and technology node will determine the degradation a chip experiences. Since temperature is dependent on many circuit parameters (e.g., threshold voltage), the degradation will likely result in a different temperature profile over time. A recycled chip which has experienced prior degradation may therefore exhibit thermal characteristics that differ from a fresh (unused) chip. An incorrect temperature profile may also result from remarked, cloned, overproduced, or defective counterfeit types since they may also possess different electrical parameters which influence thermal behavior.

3.5.1.7 Delay Defects (EP7)

This defect refers to additional delays introduced to circuit paths due to imperfections in manufacturing process or degradation from prior use. If a non-critical path of a circuit becomes critical, the circuit may fail during the operation.

3.5.2 Manufacturing Defects

These defects are classified into three categories: process, material, and packaging defects (see Fig. 3.37). Process defects come from the photolithography and etching processes during fabrication. The misalignment of photo-masks and over/under etching results in process defects. The defects related to "material" are the defects that arise from the impurities within the silicon or oxide layers. Crystalline defects in silicon changes the generation-recombination of carriers and eventually results in the failure of the device. Crystal imperfections, surface impurities, and improper materials come under this category. We will describe these defects in more detail below.

As these defects defect arise from the manufacturing process, they may potentially be present in (i) out-of-spec/defective components when an untrusted entity supply the defective components in the market, and (ii) overproduced and cloned components when the untrusted foundry source the components without proper tests. We will probably not find these defects in recycled and remarked components as they are assumed to be tested properly when they originally entered into the supply chain.

3.5.2.1 Missing Contact Windows (EM1)

The misalignment of photo-masks and over or under etching results in missing contact windows. The gate of a transistor will float when the metal-to-polysilicon windows of that transistor are missing.

3.5.2.2 Misaligned Window (EM2)

Misaligned window occurs when the photo-masks are misaligned. It affects the current carrying capability and may form parasitic transistors.

3.5.2.3 Oxide Break-Down (EM3)

The MOS devices with very thin oxide layers are generally subjected to a very high electric field. This imperfection occurred during the manufacturing process and remains in the device if it is not tested properly. This defect may potentially be observed in overproduced, out-of-spec, or defective counterfeit types.

3.5.2.4 Parasitic Transistors (EM4)

Due to the over-etched and misaligned contact windows, parasitic transistor actions may occur between adjacent devices. Electric charge buildup takes place between the two adjacent diffusion regions under an electric field that is sufficiently strong to revert the layer to a conducting channel, resulting in the device's failure.

3.5.2.5 Fine Cracks (EM5)

Fine cracks may occur during the mishandling of dies at various stages in the fabrication process.

3.5.2.6 Crystal Imperfection (EM6)

Crystalline defects in silicon changes the generation-recombination of carriers and eventually results in the failure of the device.

3.5.2.7 Surface Impurities (EM7)

These defects arise from impurities within the silicon or oxide layers.

3.5.2.8 Improper Materials (Seals, Epoxies, Dielectrics, etc.) (EM8)

These types of defects arise from improper materials being present in seals, epoxies, dielectrics, etc.

3.5.2.9 Surface Passivation and Corrosion (EM9)

The passivation layer provides some form of protection for the die. Failure occurs when corrosion causes cracks or pin-holes to form in the passivation layer. In addition to that, the aluminum layer can easily be contaminated and corroded by the presence of sodium and chloride, and it can potentially result in the formation of an opening.

3.5.2.10 Mechanical Interfaces: Intermetallic Growth (EM10)

Intermetallic growth in the bond and other mechanical interfaces results from the metal impurities and causes the device to fail.

3.5.2.11 Mechanical Interfaces: Fatigue (EM11)

Fatigue in the bond and other mechanical interfaces results from the temperature and causes the device to fail.

3.6 Summary

In this chapter, we presented numerous defects that can be evident in counterfeit components. In order to deduce that a particular IC is counterfeit, the presence of one or more of these defects can be attributed. Defects are varied across different types of ICs. Evidence of defects may manifest themselves in the package, silicon die, bond wires or any other IC features. Further, as counterfeiting grows in number and complexity, the quantity and intricacy of these defects become even more challenging to address. Nonetheless, this chapter has made an attempt to provide an exhaustive taxonomy of counterfeit defects that have been noted thus far.

Defects were classified into four broad categories: procedural, mechanical, environmental and electrical, depending on where and what kind of defects was detected on suspect ICs. Under procedural defects, errors or deviations can be found in the packaging. This type of defect can be detected easily when it is possible to contact the original component manufacturer and verify the authenticity of the lot information. Mechanical defects are defects found in the physical make of the IC. They may include residual marks, wrong materials, color variations and other such aberrant physical features that are usually not found in authentic components. These defects, as noted in the chapter, may be found in a variety of locations on an IC such as its leads, packages, bond wires and even on the die itself. Environmental variations include evidence of contamination or corrosion in the package or the leads, which indicate that the components could potentially be recycled. Most of these physical defects commonly arise through the process of recycling and remarking, where used and defective ICs are subjected to often crude processes and environmental extremities, which create these counterfeit defects. Lastly, electrical defects were introduced in which suspect ICs could show parametric defects based on leakage current, temperature profile etc. or manufacturing defects such as crystalline imperfections that arise during the fabrication process, all of which are detected by comparing these possible defective ICs to ICs that are verified to be genuine.

The goal for counterfeit detection is to be able to identify as many of these defects as possible, so that a part can be deemed counterfeit with high certainty (see Chap. 6 for metrics on counterfeit detection). It is to be stressed that the list of defects presented in this chapter is in no way absolute. As the nature and scope of counterfeiting grows, newer, finer defects may arise and they must be identified and added to test metrics for counterfeit detection.

References

1. U. Guin, D. DiMase, M. Tehranipoor, A comprehensive framework for counterfeit defect coverage analysis and detection assessment. J. Electron. Test. **30**(1), 25–40 (2014)
2. U. Guin, D. DiMase, M. Tehranipoor, Counterfeit integrated circuits: detection, avoidance, and the challenges ahead. J. Electron. Test. **30**(1), 9–23 (2014)
3. U. Guin, M. Tehranipoor, On selection of counterfeit IC detection methods, in *IEEE North Atlantic Test Workshop (NATW)*, (2013)
4. Intel, Shipping and Transport Media, (2004) http://www.intel.com/content/dam/www/public/us/en/documents/packaging-databooks/packaging-chapter-10-databook.pdf
5. IDEA, Acceptability of electronic components distributed in the open market, (2011) http://www.idofea.org/products/118-idea-std-1010b
6. Departnent of Defense, Performance Specification: Hybrid Microcircuits, General Specification For, (2009), http://www.dscc.dla.mil/Downloads/MilSpec/Docs/MIL-PRF-38534/prf38534.pdf
7. Texas Instruments, Device Marking Conventions, (2005), http://www.ti.com/lit/an/snoa039c/snoa039c.pdf
8. C. Abesamis, Counterfeit Parts Inspection and Detection, http://www.erai.com/CustomUploads/conference/2013/PDF/CPAT_INSPECTION.pdf
9. CTI, Counterfeit Examples: Electronic Components, (2013), http://www.cti-us.com/pdf/CCAP-101InspectExamplesA6.pdf
10. Intel, Ball Grid Array (BGA) Packaging, (2000) http://www.intel.com/content/dam/www/public/us/en/documents/packaging-databooks/packaging-chapter-14-databook.pdf
11. D. Ross, J. Roman, E. Ito, Choosing the right material for RF packaging (2007) http://www2.electronicproducts.com/Choosing_the_right_material_for_RF_packaging-article-farcrjrticona-nov2007-html.aspx
12. P. Bereznycky, Ceramic to Plastic Packaging (2010) http://www.navyb2pcoe.org/pdf/wiki/Empfasis%20RD%20-%20Ceramic%20to%20Plastic%20Packaging.pdf
13. M. Marshall, Best Detection Methods for Counterfeit Components, (2011), http://www.smta.org/chapters/files/SMTA_Great_Lakes_Chapter_Counterfeit_Components_-Integra_Mark_Marshall_(4-11_General)_handout_2.pdf
14. M. Bushnell, V. Agrawal, *Essentials of Electronic Testing for Digital, Memory, and Mixed-Signal VLSI Circuits* (Springer, New York, 2000)
15. M. Howes, D.V. Morgan, *Reliability and Degradation of Semiconductor Devices and Circuits* (Wiley, Chichester, 1981)
16. M. Alam, S. Mahapatra, A comprehensive model of pmos nbti degradation. Microelectron. Reliab. **45**(1), 71–81 (2005)
17. S. Bhardwaj, W. Wang, R. Vattikonda, Y. Cao, S. Vrudhula, Predictive modeling of the nbti effect for reliable design, in *Proc. of IEEE on Custom Integrated Circuits Conference*, (2006), pp. 189–192
18. V. Reddy, A. Krishnan, A. Marshall, J. Rodriguez, S. Natarajan, T. Rost, S. Krishnan, Impact of negative bias temperature instability on digital circuit reliability, in *Proc. on Reliability Physics*, (2002), pp. 248–254
19. D. K. Schroder, J. A. Babcock, Negative bias temperature instability: Road to cross in deep submicron silicon semiconductor manufacturing. Appl. Phys. **94**(1), 1–18 (2003)

20. W. Wang, V. Reddy, A. Krishnan, R. Vattikonda, S. Krishnan, Y. Cao, Compact modeling and simulation of circuit reliability for 65-nm cmos technology. IEEE Trans. Device Mater. Reliab. **7**(4), 509–517 (2007)
21. K.-L. Chen, S. Saller, I. Groves, D. Scott, Reliability effects on mos transistors due to hot-carrier injection. IEEE Trans. Device Mater. Reliab. **32**(2), 386–393 (1985)
22. S. Mahapatra, D. Saha, D. Varghese, P. Kumar, On the generation and recovery of interface traps in mosfets subjected to nbti, fn, and hci stress. IEEE Trans. Device Mater. Reliab **53**(7), 1583–1592 (2006)
23. J. McPherson, Reliability challenges for 45 nm and beyond, in *Proc. of ACM/IEEE on Design Automation Conference*, (2006), pp. 176–181
24. G. Groeseneken, R. Degraeve, T. Nigam, G. Van den Bosch, H. E. Maes, Hot carrier degradation and time-dependent dielectric breakdown in oxides. Microelectron. Eng. **49**(1–2), pp. 27–40 (1999)
25. J. Stathis, Physical and predictive models of ultrathin oxide reliability in cmos devices and circuits. IEEE Trans. Device Mater. Reliab. **1**(1), 43–59 (2001)
26. J. Black, Electromigration—a brief survey and some recent results. IEEE Trans. Electron Devices **16**(4), 338–347 (1969)
27. H. Iwai, Roadmap for 22 nm and beyond (invited paper). Microelectron. Eng. **86**(7–9), 1520–1528 (2009). doi:10.1016/j.mee.2009.03.129
28. N. Kim, T. Austin, D. Baauw, T. Mudge, K. Flautner, J. Hu, M. Irwin, M. Kandemir, V. Narayanan, Leakage current: Moore's law meets static power. Computer **36**(12), 68–75 (2003)

Chapter 4
Physical Tests for Counterfeit Detection

With the proliferation of counterfeit components in the electronic component supply chain over the last decade, it has become imperative that manufacturers, distributors, and users of electronic components inspect all incoming electronic components for authenticity, especially for those parts that will be used in critical systems, infrastructures, and applications (aerospace, military, medical, transportation, etc.). The risk of using a tampered, unreliable, or untrustworthy counterfeit component in such systems can be catastrophic (i.e., life-or-death). In addition, critical systems are often composed of older and obsolete electronic components which are no longer available from the original component manufacturers (OCMs) or OCM-authorized distributors. Following the laws of supply and demand, obsolete components are therefore more expensive, thus increasing the financial incentive for counterfeiters to source fakes of such parts.

Several test methods have been developed to distinguish counterfeit from authentic components. The goal of each test is to spot one or more of the counterfeit defects discussed in the previous chapter. Guidance, requirements, and procedures for carrying out such tests have been outlined by several standards [1–4], but it is very important that the community understands the challenges and limitations associated with these tests as well.

In this chapter, we first present a detailed taxonomy of counterfeit detection methods which can be broadly classified into two distinct categories—physical tests and electrical tests. This chapter is devoted to the physical test category. Physical test are focused on capturing the defects related to the exterior, interior, and materials of a components package, leads, and die. The physical tests discussed here range from simple external visual inspection (EVI) to more advanced imaging techniques that rely on scanning electronic microscopes (SEMs) and X-ray tomography. We shall discuss all aspects of these methods including their objectives, required equipment, as well as their challenges and limitations. The electrical test category will be discussed in Chap. 5.

M. Tehranipoor et al., *Counterfeit Integrated Circuits*,
DOI 10.1007/978-3-319-11824-6_4

4.1 Taxonomy of Counterfeit Detection Methods

Figure 4.1 shows a taxonomy of counterfeit detection methods. Physical tests are performed to examine the physical and chemical/material properties of the component's package, leads and die of a component in order to detect procedural, mechanical, and environmental counterfeit defects (see Chap. 3). When an order is received, physical inspection is the first set of tests conducted, in which the ordered component in question is tested for possible evidence of counterfeiting. As part of the physical inspection procedure, the component is thoroughly inspected using imaging techniques of the exterior and interior. The exterior part of the package and leads of the component are analyzed using exterior tests. For example, the physical dimensions of the components are measured either by hand-held or automated test equipment. Any abnormal deviation of measurement from the specification sheet indicates that the component may be counterfeit.

The chemical composition of the component is verified using material analysis. Defects such as wrong materials, contamination, oxidation of leads and packages, etc., can be detected. There are several tests that can perform material analysis (e.g., XRF, EDS, FTIR, etc.).

The internal structures—die, and bond wires—of the components may be inspected by delid/decapsulation or X-ray imaging. There are three mainstream methods commercially available for decapsulation: chemical, mechanical or laser-based solutions. Chemical decapsulation involves etching away the package with an acid solution. Newer laser-based techniques can remove an area of the package. Mechanical decapsulation involves grinding the part until the die is exposed. Once the part has been decapsulated and the required structures exposed, the interior tests need to be performed. These may include observation of the presence of an authentic die, gross cracks on the die, delamination, any damage on the die, die marking, missing or broken bond wires, reworked bonds, bond-pull strength, etc.

Electrical tests are the only way to determine the correct functionality of a component. These tests are a very efficient and non-destructive way of detecting counterfeit components. The majority of defects listed under electrical category (see Fig. 3.1) can be effectively detected by electrical tests. In addition, die and bond wire related defects may also be detected by these tests. The major advantage of introducing electrical tests in to a test plan is that they can identify cloned, out-of-spec/defective, and overproduced components along with the recycled and remarked components, as most of the electrical defects may be present in those components (see Chap. 3, Sect. 3.5.2). We will introduce these tests in detail in Chap. 5.

The taxonomy of counterfeit detection methods was introduced in [5–9]. However, in this chapter, we modify the taxonomy to be in line with the counterfeit detection standards [1–4].

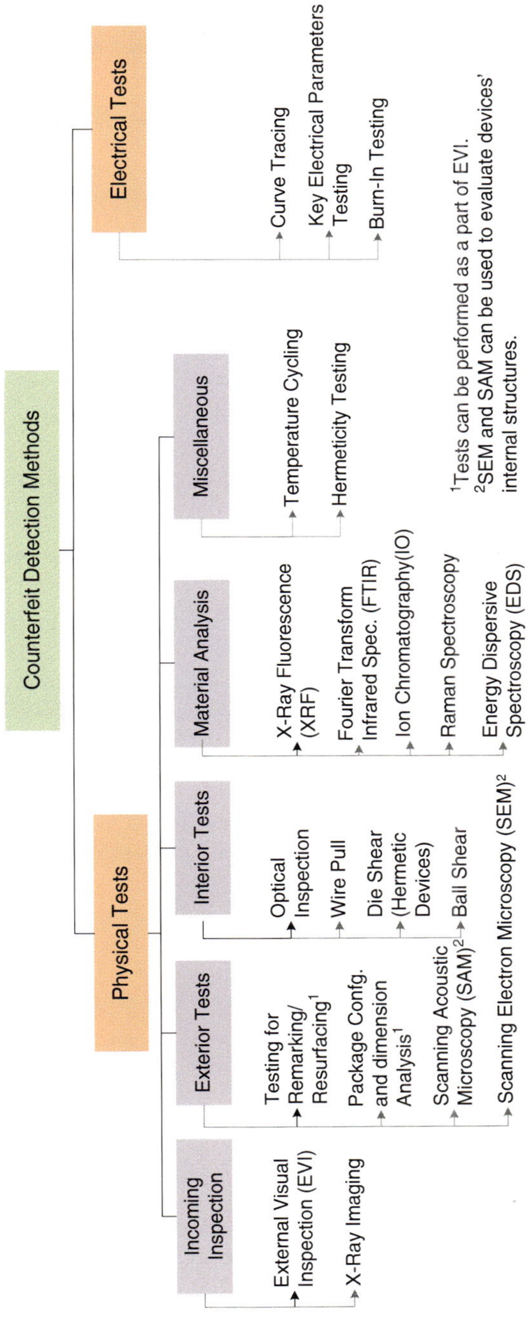

Fig. 4.1 A taxonomy of counterfeit detection methods

4.2 Physical Inspection

The first set of tests that is to be applied to defect counterfeit components are physical inspections. The physical and chemical properties of components' package, leads and die are thoroughly inspected by physical inspection, as we described earlier. Some of the physical inspections, e.g., EVI, X-Ray imaging, etc., are easy to implement whereas some of them require custom-designed expensive equipment, e.g., SEM, SAM, etc., and skilled operators to perform the tests. Sometimes subject matter experts (SMEs) are necessary to interpret the test results. In the following subsections, we will describe some effective physical inspection methods to detect counterfeit components.

4.2.1 External Visual Inspection (EVI)

External visual inspection is the first test usually performed on all the components. The inspection is carried out in multiple steps.

General EVI The conditions for the packaging and shipping materials are verified by general EVI. Along with this, some of the physical attributes of the components are verified. However, the components are not removed from the tape or reel. A low power microscope may be required to inspect the parts (generally less than 10X magnification). All the components are handled with proper precautions as described in ANSI/ESD S20.20 for ESD components and IPC/JEDEC J-STD-20 and J-STD-033 for moisture sensitive components. The inspection generally verifies the following:

i. Packaging: The verification of packaging is performed and compared with the original component manufacturer (OCM) packaging. Any damage on the packaging is also carefully inspected.
ii. Documents: The documents received with the packaging, external and internal shipping labels are verified. OCM's logs, shipping origin, certificate of conformance (CoC) are scrutinized and verified with the OCM.
iii. Protection: The electrostatic discharge (ESD) sensitive components shall be shipped in accordance with EIA standard 541's electrostatic discharge (ESD) requirements. Similarly, the moisture sensitive components shall be packaged in accordance with the requirements of IPC/JEDEC J-STD-020C. A thorough inspection of ESD sensitive bags, humidity indicator cards (HIC), and moisture sensitive bags are performed. All the components are also handled with proper care so that no components get damaged.
iv. Orientation: The orientations of all the components in a packaging tape or reel are checked. In an authentic lot, the components are oriented in the same direction. For example, the placement of the pin 1 marker of a component in tape is towards the operator. Different orientations in a package indicate

that some or all components may have been replaced with their counterfeit counterparts. Further verification is needed to authenticate such component.

v. Invalid Marking: Validity of the marking (see the defect *invalid marking* in Chap. 3, Sect. 3.2) on the component is checked. For example, the date code need to be checked for all the components. Old components are sometimes remarked with a current date code to make them look like newly manufactured components.

vi. Gross Visual Anomalies: Gross visual anomalies are also inspected. For example, damaged leads or packages, sanding or grinding marks on the package, color variations of leads, color variations of packages for different components in a same lot, etc., are audited.

General EVI is the only physical test that is applied to all the components under authentication due to its speed and low cost. After this inspection, sampling is typically performed where a few components are randomly selected from a lot for further tests to validate the authenticity of the purchased parts.

Detailed EVI Detailed EVI is performed on the sampled components with a microscope with 10X–40X magnification. Higher magnification (up to 100X) may be necessary to spot some of the finer defects. Figure 4.2 illustrates a set up

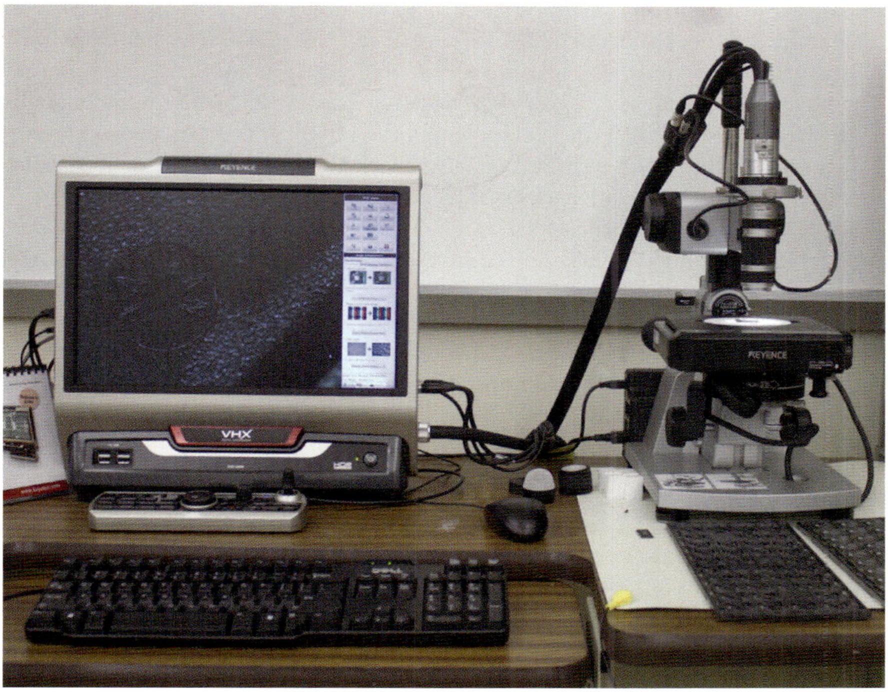

Fig. 4.2 An inspection set up for detailed EVI (CHASE Center, University of Connecticut)

for detailed visual inspection. The setup consists of a Keyence VHX-2000 digital microscope [10] which is a versatile imaging and measurement system with 0.1X–5000X magnification. It can capture full focused images, observe objects from any angle, and even visualize surfaces in three dimensions (3D).

The handling of components is of particular concern during imaging. All the sampled components should be handled in accordance with the requirements of ANSI/ESD S20.20 for ESD components and IPC/JEDEC J-STD-20 and J-STD-033 for moisture sensitive components. Detailed EVI generally performed to validate the following,

i. Leads, Balls, and Columns: The inspection of leads for through-hole components, and balls and columns for surface mount components are carefully performed. This detailed inspection includes the search for—(a) scratches on the leads, (b) bent, broken, and missing leads, (c) residual material on the leads, (d) replating on the leads, (e) straightness, pitch, and separation of the leads, (f) tooling marks on the leads, (g) misaligned and missing balls and columns, and (h) distorted and nonuniform balls and columns.
ii. Package: The package of an IC can reveal significant information about the authenticity of a chip. The detailed inspection of the package of a component is performed in this step. The authenticity of the marking is verified. For example, invalid date, lot, or country codes are checked. If the package exhibits any external sanding or grinding marks, it has likely been remarked. The labels that are on the package should be permanent and clean. The markings on a counterfeit part may be crooked, uneven, or sloppy. An imprecise laser used by a counterfeiter may also hover over spots for too long and cause burn marks on the package. Ghost markings, color variations, improper textures, and extraneous marking on the package surface are also checked. The package is also inspected carefully to find any damage caused by improper handling.

    The location of the pin 1 marker and mold markers in the package should be identical for components from the same lot. These marker cavities should also be free of scratches and extra materials. The edges around the mold markers should be sharp and precise. If the edges seem to be rounded down, it may be an indication that the package was sanded down for remarking. The dimensions of the package are also verified against datasheet specifications.

Testing for Remarking and Resurfacing The purpose of this test is to determine the quality of the surface coating and marking on the package of a component in order to determine whether the component is blacktopped and remarked. It is absolutely necessary to evaluate the package surface to find the defects and anomalies related to the marking and package surface. Acetone is commonly used to determine if the component is remarked or not. Some harsher solvents (Dynasolv 711 or 750) may also be necessary for this test. If the surface or marking color changes, the component is assumed to be counterfeit. The detailed test procedure can be found in method 2015 of MIL-STD-883.

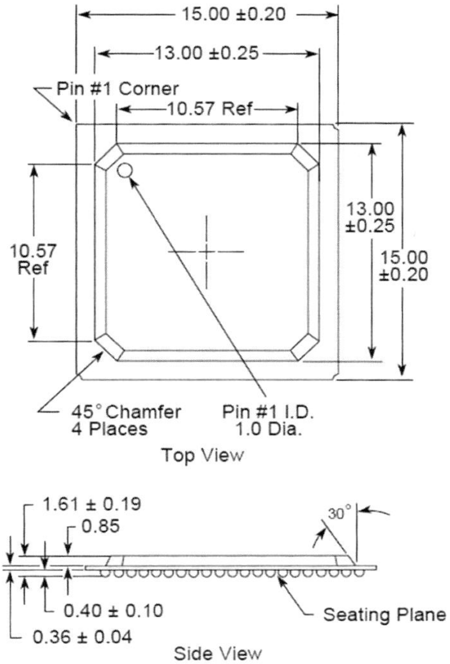

Fig. 4.3 Dimensions of Intel's 15 mm PBGA outline drawing [11]

Package Configuration and Dimension Analysis In this phase, the physical dimensions of the components are measured either by hand-held or automated test equipment. Figure 4.3 shows the dimensions of Intel's 15 mm Plastic Ball Grid Array (PBGA) outline drawing. Any abnormal deviation of measurement from the specification sheet must be recorded and the component shall require further analysis.

4.2.2 X-Ray Imaging

X-Ray imaging is a method of inspecting the internal structure of a component without performing decapsulation. There are typically two types of X-Ray imaging systems—film X-Ray and real-time X-Ray systems. In film X-Ray systems, the images are formed on a radiographic film, whereas in real-time, or digital X-Ray systems, a digital image is formed by digital sensors. Figure 4.4 shows a real time X-Ray imaging system, Zeiss Versa 510 (which has a maximum power of 160 kV for its source), at the University of Connecticut which has been used to acquire the structural information of different ICs. The ZEISS Xradia 510 Versa 3D X-ray Microscopes (XRM) [12] provides a unique RaaD (resolution at a distance)

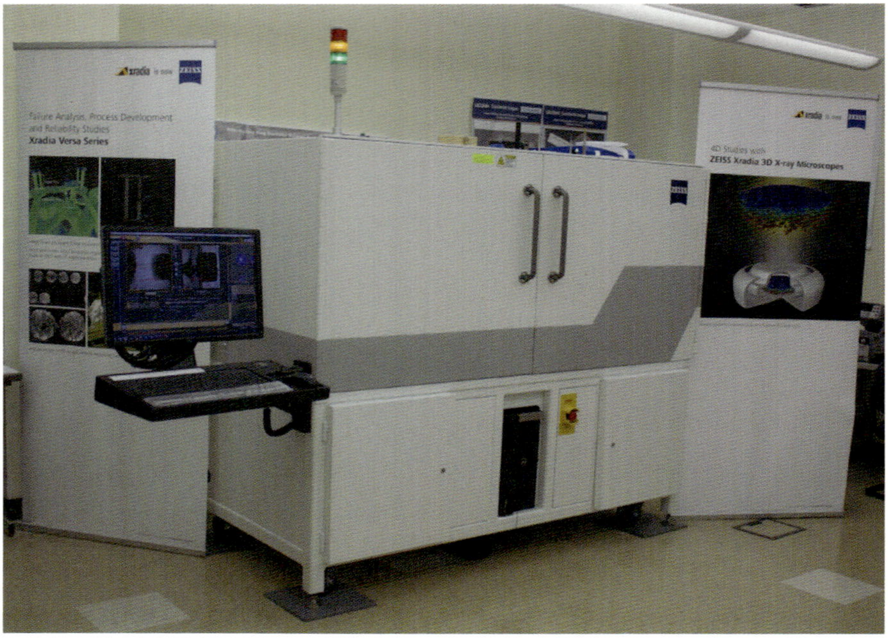

Fig. 4.4 Zeiss Xradia 510 Versa: X-ray imaging system (CHASE Center, University of Connecticut)

capability that breaks the one micron resolution barrier for samples from mm to cm. The system is used to capture two dimensional (2D) projections and X-Ray computed tomography to observe the detailed three dimensional (3D) internal structure of an IC.

X-Ray Computed Tomography (CT) is a non-invasive imaging technique that makes it possible to visualize the 3D internal structure of an IC. Multiple 2D projections are collected from many different angles with different magnifications depending on the quality needed for the final image. These 2D images are then stacked up and a 3D image is reconstructed by using mathematical algorithms, direct Fourier transform and center slice theory [13]. In order to produce a successful reconstruction of a 3D image of an IC, the following parameters should be optimized carefully: source/detector distance to object, source power, detector objective, filter, exposure time, number of projections, center shift, and beam hardening. We will describe these parameters in detail in Chap. 7. Interested readers are also suggested to read [14] for further details. Figure 4.5 shows 3D images showing the internal structure of components.

A wide variety of defects, internal to ICs, can be detected by X-Ray imaging. These defects are broadly classified into two categories—the defects related to die and those related to bond wires. Die related defects include missing die, wrong die, and gross cracks. Missing wires, broken wires, reworked bonds, poor/inconsistent

Fig. 4.5 3D Structure of an Intel TB28F400B5-B80 flash memory

lead dress/frame, etc. are some common bond wire related defects that can be detected by X-Ray imaging. Figure 4.6a and b shows two different lead frame structures for two Intel TB28F400B5-B80 flash memories. The die is rotated in Fig. 4.6b which is a clear indication of a wrong die. There are no bond wires inside the package that is shown in Fig. 4.6c. Broken bond wires are presented in Fig. 4.6d.

4.2.3 Delid/Decapsulation

To fully inspect the internal structure of an IC, it is necessary to remove the outer package carefully without damaging the die. Die markings are quite important in authenticating the company name, logo, date of manufacturing, mask numbers, device specification, etc. The major functional blocks are visible after decapsulating of an IC. The die sizes may vary and generally become smaller when the company moves to a newer technology node. To authenticate an IC, it is extremely important to observe these internal parameters. Note that this is a destructive test and would normally be performed on a sampling basis (e.g., a sampling plan can be found in AS6171 [2]). After decapsulation, the internal structures are thoroughly observed (see MIL-STD-883 [15] method 2013). The detailed descriptions can be found in MIL-STD-883 [15] method 5009 for the microcircuits, and MIL-STD-1580 [16] for electronic, electromagnetic, and electromechanical Parts.

4.2.4 Scanning Acoustic Microscopy (SAM)

SAM is one of the most efficient ways of studying the structure of a component without damaging it. This technology utilizes the reflection and transmission of ultrasound waves to generate an image of the component based on its acoustic

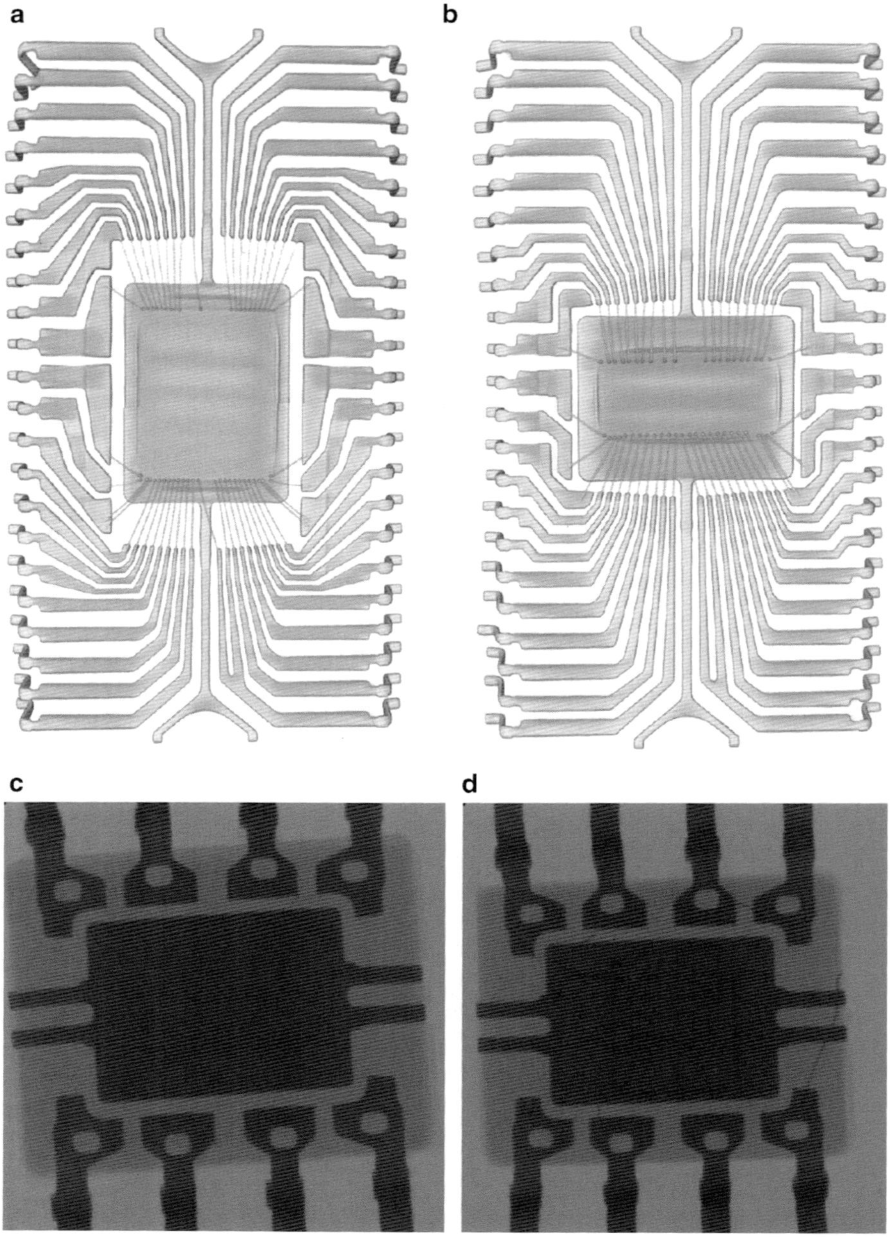

Fig. 4.6 Counterfeit defects detected by X-ray imaging (*source*: Honeywell) (a), (b), (c) Missing bond wires, and (d) Broken bond wires

impedance at various depths. The component under test is submerged in either deionized water or isopropyl alcohol (IPA), which is used as a medium. Since air will have a much different acoustic impedance than any of the parts mediums, that section will appear much darker on the image produced. The resolution of SAM depends on the transducer frequencies. Lower frequencies provide higher penetration through the component at the cost of lower spatial resolution. SAM is very useful in detecting delamination, or, die attachment to the package. It can also detect the cracks and voids in the die and anomalies in the bond wires.

4.2.5 Scanning Electron Microscopy (SEM)

SEM is a method of generating an image with a superfine resolution by using a focused electron beam. The image is formed by scanning the entire target area of the sample surface, which produces various signals that provide compositional and surface topography information. SEM consists of two major components—the electron column and a control console. The column generates a focused electron beam for scanning the surface and the control console displays the image. When the high energetic electron beam interacts with the sample, it generates a secondary emission of back scattered electrons and X-rays. An electron detector detects these secondary electrons and an image is formed. The components under test can be observed in high vacuum and low vacuum environment depending on the required information from it. Although the component is located inside a vacuum environment, tilting the stage is possible and the stage can be also heated during the microscopy. A detailed description can be found in [17].

Figure 4.7 shows an FEI Quanta 250 FEG field emission scanning electron microscope [18], located at the Center for Clean Energy Engineering at the University of Connecticut, with versatile, high resolution low vacuum capabilities. It provides a pre-aligned electron optical column for high resolution and beam stability.

SEM is very useful for detecting many defects and anomalies present in counterfeit components. The detailed inspection of a component is performed in three different ways.

i. Inspection of Leads: The surface morphology of the leads of a component can be inspected. Reworked leads, wrong materials, scratches, oxidation, etc. can easily be detected.
ii. Inspection of Package: A wide variety of defects related to the package can be detected. For example, sanding and grinding marks, ghost markings, burned markings, improper texture, oxidation, corrosion, contamination, etc. are some of the defects that can be detected by SEM. Figure 4.8 shows the marking differences of the package. The images were taken in a low vacuum medium without applying any conductive coating on the marking. The reader can easily differentiate the two markings. Note that these differences were not visible via Detailed EVI.

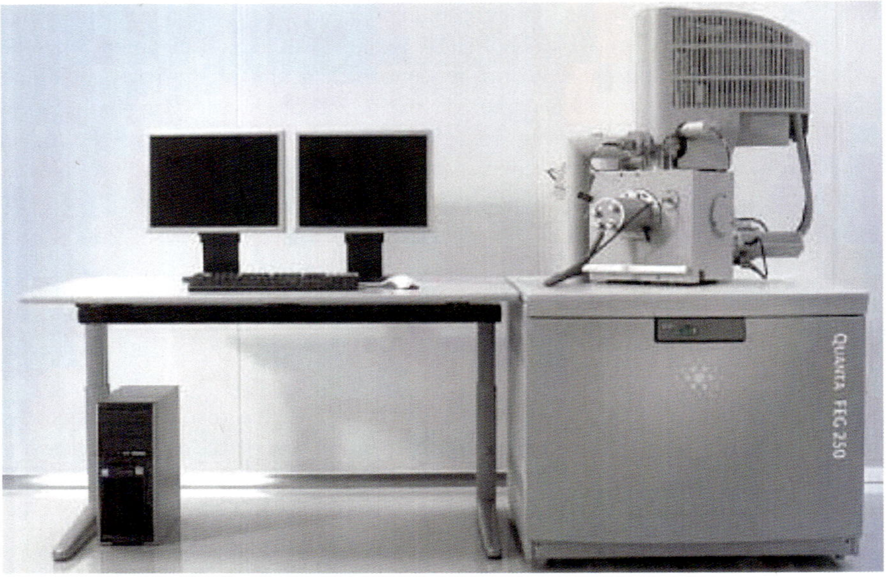

Fig. 4.7 FEI Quanta 250 FEG field emission scanning electron microscope (center for Clean Energy Engineering, University of Connecticut)

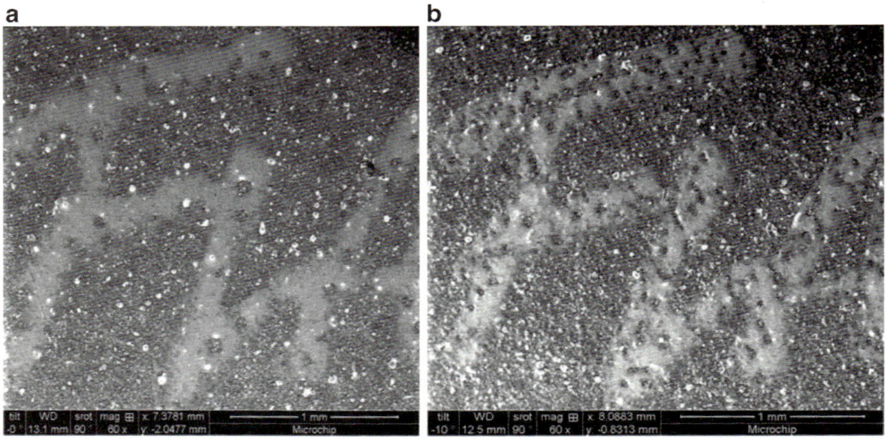

Fig. 4.8 SEM images of the marking defects (**a**) Smooth and thinner marking (**b**) Perforated marking

iii. Inspection of Die: Inspection of the die requires decapsulation of a component. The markings on the die, reworked bond wires, die damage, extraneous marks, etc. can be inspected at high resolution by SEM.

Even though SEM is very useful for detecting defects, the practicality of SEM is limited due to the long test time. *Sometimes it requires several hours to inspect a single component in detail.*

4.2.6 X-Ray Fluorescence (XRF) Spectroscopy

XRF is a nondestructive method for material analysis. The emission characteristics of a material are observed after heavy bombardment with high-energy X-Rays. When the X-Ray hits the surface of a material, the outer electrons obtain enough energy (ionization potential) to reach unstable higher outer orbits. The emission of radiation occurs when these high-energy electrons settle down to their original ground state. Each element produces a unique peak in the spectrum. A unique fingerprint is generated from the package of a component by XRF Spectroscopy. A decision about a component's authenticity can be made upon comparison with a golden sample (if available). There are several X-Ray fluorescence spectrometers with an automated sample feed that are available for material analysis.

4.2.7 Fourier Transform Infrared (FTIR) Spectroscopy

FTIR is based on infrared (IR) spectroscopy. A part of IR radiation is absorbed by the material under test and the other part is transmitted through it. The spectrum for molecular absorption and transmission is observed from the resultant IR radiation. The unique molecular fingerprint, created by FTIR, is compared to the fingerprint of the golden model for material comparison. FTIR is used to authenticate both organic and inorganic materials of a component. It is used to verify—(i) polymer, coating, etc., of the package, (ii) residual foreign materials from the sand blasting process used to remove the old markings, and (iii) residuals from chemically etched package.

4.2.8 Energy Dispersive Spectroscopy (EDS)

EDS is used to chemically characterize a component using X-ray excitation. A high-energy beam of charged particles is bombarded on the surface, and the emitted X-ray spectrum is captured by an X-ray detector to form the EDS spectrum. A unique fingerprint of X-ray spectrum is generated as each element or material has a unique atomic structure that allows a unique set of peaks on its X-ray emission spectrum. This is the basis for elemental analysis using EDS. FEI Quanta FEG 250 (see Fig. 4.7) is a combined machine which has both SEM and EDS tools on it.

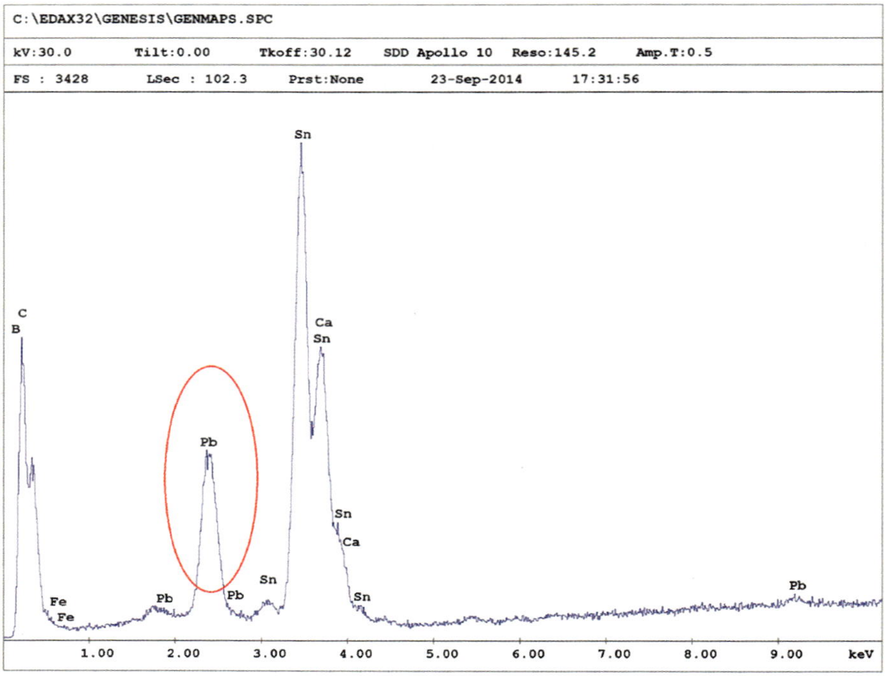

Fig. 4.9 EDS spectrum of the lead of an IDT Dual-Port Static RAM

Figure 4.9 shows the EDS spectrum of one the lead of an IDT Dual-Port Static RAM (*IDT* 7025*S*45*PF*). The authentic component is ROHS Compliant and should be free from *Pb*, which is usually restricted on IC components for environmental safety reasons. However, *Pb* is detected signifying that this must be a counterfeit component.

4.2.9 Temperature Cycling

Temperature cycling is one of the major tests that can be implemented to detect recycled ICs, as it determines the resistance of an IC to the extreme (very high and very low) temperatures. This test provides the assessment of package quality. Since temperature cycling is destructive, only a small batch of ICs can be used for testing.

Table 4.1 shows the test conditions for temperature cycling. As specified in MIL-STD-883 method 1010.7, this test shall be conducted for at least ten cycles using the test condition C specified in Table 4.1, whereas one cycle consists of the sequence of step 1 and step 2, and vice versa. Figure 4.10 shows a typical temperature profile during temperature cycling [19]. The component under test is

Table 4.1 Test conditions for temperature cycling [15]

Step	Minutes	Temperature (°C)											
		A		B		C		D		E		F	
1	$\geq$ 10	−55	+0	−55	+0	−65	+0	−65	+0	−65	+0	−65	+0
Cold			−10		−10		−10		−10		−10		−10
2	$\geq$ 10	85	+10	125	+15	150	+15	200	+15	300	+15	175	+15
Hot			−0		−0		−0		−0		−0		−0

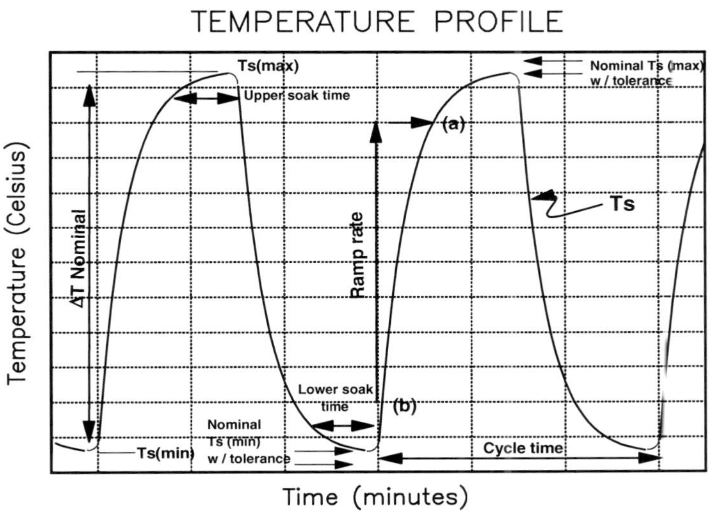

Fig. 4.10 A typical temperature profile during temperature cycling [19]

placed in a stationary chamber and is heated or cooled by using hot or cold air. The following parameters need to be controlled properly during temperature cycling.

i. *Soak Temperature* is temperature range ($Ts(max)$ and $Ts(min)$) specified in Table 4.1.
ii. *Nominal* ΔT is the difference between nominal Ts(max) and nominal Ts(min).
iii. *Soak Time* is the total time the sample temperature is within a specified range of each nominal $Ts(max)$ and nominal $Ts(min)$.
iv. *Cycle Time* is the time interval between the two consecutive high-temperature extremes or the two consecutive low-temperature extremes.
v. *Ramp Rate* is the rate of temperature increase or decrease per unit of time.

The defects or anomalies generated from the recycling process related to the package (package damage), bond wires (bond pull strength, poor connection, etc.), or die (delamination, gross cracks, die damage, etc.) can be detected by this test. A detailed description can be found in standard MIL-STD-883 method 1010.7

[15] and JESD22-A104D [19]. Along with temperature cycling, thermal shock, described in MIL-STD-883 method 1011.9, can be implemented to increase the test coverage, and to help detect recycled ICs.

4.2.10 Hermetic Seal Test

Seal test is performed to determine the quality of a components seal. The seal of a component protects the entrance of damaging contaminants which will reduce its effective life. The seal of a hermetically sealed part ensures its correct operation in the environment that it was designed for. A break in this seal leads to the failure of the component. Seal test is designed to detect leaks resulting from the use of inferior sealing materials, the manufacturing processes making the seal, or the counterfeiters damaging the seal during the recycling process.

The detailed descriptions for the seal test can be found in (i) MIL-STD-883 [15] method 1014 for microcircuits, (ii) MIL-STD-750 [20] method 1071 for active discrete components, and (iii) MIL-STD-202 [21] method 112 for passive components.

4.3 Limitations and Challenges

The counterfeiting of ICs is a multifaceted and evolving problem. Counterfeiters are enriching their knowledge and technology as they mature and become more proficient in this illegal business. The test methods that are capable of detecting counterfeit ICs today may not be as effective in the near future. Thus, it is important to analyze the limitations and challenges associated with these tests. In the following, we will briefly describe these limitations and challenges.

- Counterfeit Types: The physical inspections are primarily designed to detect recycled and remarked ICs. These tests are not nearly as effective for detecting any other counterfeit (overproduced, cloned, and out-of-spec/defective) types.
- Dynamic Nature: The dynamic nature of counterfeiting makes the detection even more challenging. Counterfeiters are evolving and adapting to new ways of making more deceptive counterfeit product. Currently, detection is mostly based on inspecting the physical appearances of devices. It is hardly a matter of time before some of these test methods will be ineffective in the near future. Some defects may no longer appear in the future and will be replaced by other defects that are not yet in the taxonomies. New tests will need to be developed in order to identify and keep up with the rapid improvements made by counterfeiters.
- Destructiveness and Sampling Requirements: Most of the physical tests are destructive by nature. Sample preparation is extremely important as it directly relates to test confidence. If a few counterfeit components are mixed with a large

batch, the probability of selecting the counterfeit one for test is extremely small. If one could develop tests that are non-destructive and efficient, we can remove the need for sampling and be able to overcome this issue.

- Test Time and Cost: The test time and cost are major limiting factors in the use of physical tests for counterfeit detection. The equipment used for physical inspections (e.g., scanning electron microscopy (SEM) and acoustic microscopy (SAM)) are not custom-designed to detect counterfeit parts. It can take several hours (e.g., typically several hours for SEM analysis) to test a single component with good resolution. This is a severely limiting factor since a large number of parts need to be inspected in order to keep up with the dynamic nature of counterfeiters and new trends in counterfeiting (identify new defects, new counterfeit types, etc.). There is little work present in the literature that evaluates the effectiveness of physical inspection tests. If we are ever to decrease test time and cost, there needs to be a strong framework in place to determine which are the most effective tests, which defects are detectable by what tests, which are the most important defects, etc. In addition, test time and cost can also be reduced by designing components with counterfeit detection in mind.
- Lack of Automation and Need for Quantitative Metrics: All the tests described here are currently performed in an ad-hoc fashion with no metrics for quantifying against a set of counterfeit types, anomalies, and defects. Most of the tests are carried out without automation. The test results mostly depend on the subject matter experts (SMEs). The decision-making process is entirely dependent on the operator (or SMEs)—this is indeed error prone. A chip could be considered counterfeit in one lab while it could be marked as authentic in another lab This was proven by a test run by G-19A group, where some labs reported a chip as counterfeit and others labeled it authentic [22]. There is a substantial need to develop automated test methods that allow us to quickly and more efficiently identify counterfeit defects on a large number of parts. Such results could be used to keep track of trends in counterfeiting and stay on pace with counterfeiters. In order to reach this goal, metrics are needed to quantify (i) many of the counterfeit defects only identifiable today by SMEs and (ii) many of the test methods and equipment described above.

Preliminary work on overcoming many of the above challenges is contained in latter chapters of this book. These include assessment and optimization of test methods (Chap. 6), advanced physical tests (Chap. 7), and design-for-anti-counterfeit measures (Chaps. 9 –12).

4.4 Summary

In this chapter, we have presented a detailed taxonomy of test methods for the detection of counterfeit components. These test methods are classified into two distinct categories: physical and electrical. For this chapter, we have focused on various physical test methods. General external visual inspection (EVI) is a physical test method that can be applied to all incoming components. 2D X-Ray imaging can be applied as a test for observing the internal structure of an IC for possible die and bonding defects whereas 3D tomography can be used to generate more detailed images of the internal structure of the IC.

We have also analyzed the challenges and limitations of these tests, mainly in terms of the time and cost-constraints. EVI is fairly cheap and quick to administer as it involves the use of simple low-power microscopes for visualization of the defects. Spectroscopy methods to determine the material composition of ICs for detecting sanding residues, detailed X-ray imaging, SEM, and SAM techniques are more expensive and require extensive amounts of time and effort to prepare samples and conduct the test. In fact, all the physical tests introduced in this chapter except general EVI can only be used on a sampling basis and cannot be used to test entire lots of ICs as they are destructive in nature, costly, and time-consuming. In the absence of automation strategies, these tests can be even more tedious to conduct. Thus, in a generic counterfeit detection scheme, general EVI is performed on all ICs from a lot and then, more advanced techniques such as X-ray imaging and spectroscopy can be used to test sampled components from the lot. To combat the difficulties of physical tests, electrical tests are introduced in Chap. 5, which can be applied more efficiently to all components in a single lot.

References

1. SAE, Counterfeit electronic parts; avoidance, detection, mitigation, and disposition (2009), http://standards.sae.org/as5553/
2. SAE, Test methods standard; counterfeit electronic parts. Work in Progress, http://standards.sae.org/wip/as6171/
3. CTI, Certification for coutnerfeit components avoidance program. September 2011
4. IDEA, Acceptability of electronic components distributed in the open market, http://www.idofea.org/products/118-idea-std-1010b
5. U. Guin, M. Tehranipoor, D. DiMase, M. Megrdichian, Counterfeit IC detection and challenges ahead. ACM/SIGDA E-NEWSLETTER **43**(3), (2013)
6. U. Guin, D. DiMase, M. Tehranipoor, A comprehensive framework for counterfeit defect coverage analysis and detection assessment. J. Electron. Test. **30**(1), 25–40 (2014)
7. U. Guin, D. DiMase, M. Tehranipoor, Counterfeit integrated circuits: detection, avoidance, and the challenges ahead. J. Electron. Test. **30**(1), 9–23 (2014)
8. U. Guin, K. Huang, D. DiMase, J. Carulli, M. Tehranipoor, Y. Makris, Counterfeit integrated circuits: a rising threat in the global semiconductor supply chain. Proc. IEEE **102**(8), 1207–1228 (2014)
9. U. Guin, M. Tehranipoor, On selection of counterfeit IC detection methods, in *IEEE North Atlantic Test Workshop (NATW)*, May 2013
10. VHX-2000 series Digital Microscope [Online], Available: http://www.keyence.com/products/microscope/digital-microscope/vhx-2000/index.jsp
11. Intel, Ball Grid Array (BGA) Packaging, http://www.intel.com/content/dam/www/public/us/en/documents/packaging-databooks/packaging-chapter-14-databook.pdf
12. ZEISS Xradia 510 Versa, Submicron X-ray Imaging: Flexible Working Distance at the Highest Resolution. [Online], Available: http://www.xradia.com/versaxrm-510/

13. X. Pan, Unified reconstruction theory for diffraction tomography, with consideration of noise control. JOSA A **15**(9), 2312–2326 (1998)
14. N. Asadizanjani, S. Shahbazmohamadi, E. H. Jordan, Investigation of surface geometry change in thermal barrier coatings using computed x-ray tomography, in *38th Int'l Conf and Expo on Advanced Ceramics and Composites, ICACC 2014*
15. Department of Defense, Test Method Standard: Microcircuits (2010), http://www.landandmaritime.dla.mil/Downloads/MilSpec/Docs/MIL-STD-883/std883.pdf
16. Department of Defense, Test Method Standard: Destructive Physical Analysis for Electronic, Electromagnetic, and Electromechanical Parts (2014), http://www.landandmaritime.dla.mil/Downloads/MilSpec/Docs/MIL-STD-1580/std1580.pdf.
17. J. Goldstein, D. Newbury, D. Joy, C. Lyman, P. Echlin, E. Lifshin, L. Sawyer, J. Michael, *Scanning Electron Microscopy and X-ray Microanalysis* (Springer, New York, 2003)
18. Quanta SEM [Online], Available: http://www.fei.com/products/sem/quanta-products/?ind=MS
19. JEDEC, JESD22-A104D: Temperature Cycling, March 2009, http://www.jedec.org/sites/default/files/docs/22a104d.pdf
20. Department of Defense, Test Method Standard: Test Methods for Semiconductor Devices (2012), http://www.landandmaritime.dla.mil/Downloads/MilSpec/Docs/MIL-STD-750/std750.pdf
21. Department of Defense, Test Method Standard: Electronic and Electrical Component Parts (2013), http://www.dscc.dla.mil/Downloads/MilSpec/Docs/MIL-STD-202/std202.pdf
22. CHASE, ARO/CHASE Special Workshop on Counterfeit Electronics, January 2013, http://www.chase.uconn.edu/arochase-special-workshop-on-counterfeit-electronics.php

Chapter 5
Electrical Tests for Counterfeit Detection

The two previous chapters have discussed the defects or anomalies that appear in counterfeit components and the physical tests that can be used to detect a subset of these defects. Also highlighted were the limitations of physical tests including their high test time and cost, destructive nature, and limitation to certain defects and certain counterfeit types. Tests such as material analysis and scanning electron or acoustic microscopy require extensive sample preparations, during the course of which the component under test becomes ineligible for further use. Added to the destructive nature, most physical tests are extremely time-consuming and above all, they cannot be used to test the functionality of an IC. Since it is desirable to be able to test as many components as possible, if not all, quick and efficient test methods are required to ensure that ICs pass stringent acceptance tests and that they meet functionality, quality, authenticity and reliability requirements.

This leads us to the concept of electrical tests which can be used to detect counterfeit components in a largely nondestructive way. In contrast to physical tests, electrical tests capture the functionality of a component which can be used to detect more counterfeit defects and counterfeit types. In fact, the majority of the counterfeit defects listed under electrical category (discussed in Chap. 3), which was undetected by physical tests, can effectively be detected by electrical tests.

In this chapter, we shall focus on electrical tests. The taxonomy of electrical tests was first introduced in [1–5]. However, here we have modified the taxonomy (see Fig. 4.1) to align more closely with the current standards for counterfeit detection. In the sections below, we begin by discussing the equipment used for electrical tests followed by three tests most often recommended by the standards: curve tracing, key electrical parameters testing, and burn-in testing. To conclude the chapter, we will present the limitations and current challenges associated with these electrical tests.

M. Tehranipoor et al., *Counterfeit Integrated Circuits*,
DOI 10.1007/978-3-319-11824-6_5

5.1 Test Equipment

The equipment used by electrical tests provide electrical signals to the component under test and collect the response from the component. There are two types: (i) bench equipment is used for more specialized and unique measurements for simple components while (ii) an automatic test equipment (ATE) [6] is used for more complex and large components (FPGAs, ASICs, microprocessors, memories, etc.).

5.1.1 Bench Equipment

Bench equipment is generally used to measure electrical parameters of a component, such as voltage, current, frequency, etc. These are stand-alone measurement devices that can independently perform the measurement. Some of the bench equipment includes ammeter, ohmmeter, voltmeter, waveform generator, oscilloscope, curve tracer, network analyzer, spectrum analyzer, etc. Figure 5.1 shows a test setup, developed at the CHASE Center of the University of Connecticut, for the detection of counterfeit ICs (here, microcontrollers).

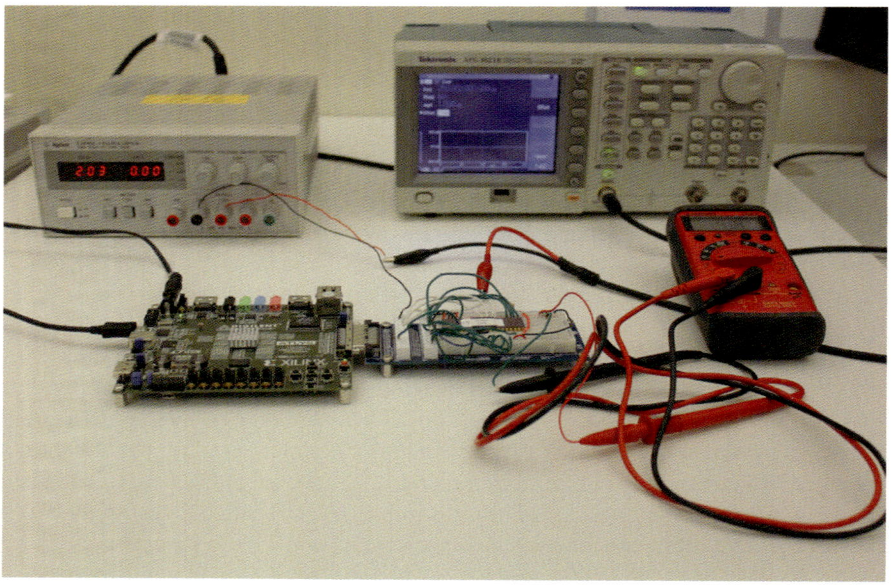

Fig. 5.1 A test setup to detect counterfeit ICs using bench equipment (CHASE Center, University of Connecticut)

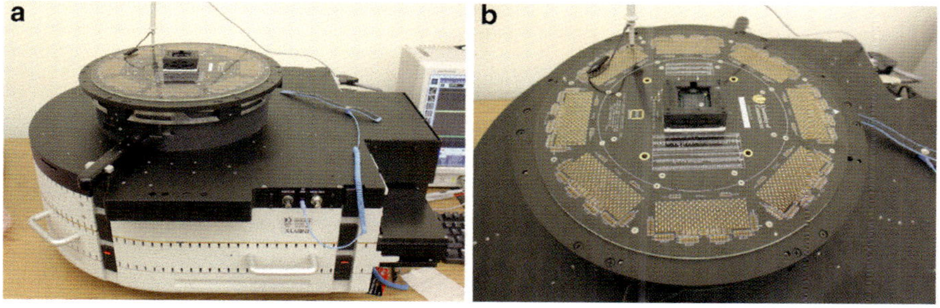

Fig. 5.2 Verigy Ocelot ZFP ATE (CHASE Center, University of Connecticut). (**a**) Verigy Ocelot ZFP, (**b**) load board

5.1.2 Automatic Test Equipment (ATE)

Automatic Test Equipment (ATE) is an instrument used to apply test patterns to an integrated circuit. It analyzes the responses from the IC, and marks it as passing if the responses match the authentic responses and failing if the responses does not match the authentic responses stored in the ATE memory. The ATE is controlled by a central UNIX work station or a Windows based personal computer. A few typical ATE vendors are Teradyne, Agilent Technologies, Advantest, Metric, Credence Systems Corporation, and National Instruments. Commercial automatic test equipment can be roughly divided into several categories based on the types of ICs they are used to test. For example, separate ATEs may be required in order to test system-on-chips (SoCs), analog ICs, mixed signal ICs, and memories.

Figure 5.2 shows a typical ATE used for the test and detection of digital counterfeit ICs. Figure 5.2b shows the loadboard used to hold the ICs. It provides an interface between the ICs under test and the ATE. Different loadboards are required to test different ICs having different pin counts and packages.

5.2 Curve Tracing

Curve tracing is gaining popularity in the detection of counterfeit components as it tests ICs non-destructively and without requiring extensive details of the ICs under test. It is not necessary to require a golden IC during authentication. In a typical curve tracer, standard voltage or current curves can be generated for any combination of pins of the ICs. These traces are formed by sweeping voltage V over a specified range and plotting the current I. The traces follow the Ohm's Law of $V = I * Z$, where Z is the impedance between the pins of an IC.

Figure 5.3 shows a typical curve tracer by National Instruments. This system can perform sweeps across different IC pins with a range of -20 to $+20$ V. A typical curve tracer generally operates in two different modes:

Fig. 5.3 A typical curve tracer by National Instruments

- Basic Curve Trace: The trace is formed from one pin to all other pins while the IC remains unpowered. In this mode, gross defects related to package (MP10, MP12, NP1, NP2, etc.), bond wire (MB1, MB2, MB3, etc.) and die (MD1, MD2, MD4, etc.) can be detected (see Chap. 3) for a description of these counterfeit defects). However, electrical, parametric and manufacturing defects cannot be detected using this approach. This mode can quickly separate a set of easy-to-detect counterfeit ICs, thus simplifying the test procedure and reducing the test time and cost needed to test these ICs with physical tests.
- Power Curve Trace: In this mode, the IC is powered on while capturing its trace. A trace is formed from a relationship of each pin to every other pin in the IC and

a fingerprint is generated. This approach can detect some of the parametric and manufacturing defects along with the defects detected by basic curve trace.

Figure 5.4 shows a typical curve trace. The trace is formed by applying voltage on the pins of a high-performance, EEPROM-based programmable logic device (EPM7096QC100-15). Figure 5.4a, b show the traces for the good pins and failed pins respectively. Table 5.1 shows the voltage and current for a few of the pins from the EPM7096QC100-15 programmable logic device. For a failed pin, the current is much higher for a small input voltage (e.g., −0.605 mA for −0.094 V and 0.592 mA for 0.086 V at pin number 13) which may possibly resulted from a resistive contact between the failed pin and the ground.

A variety of defects in ICs can be detected from curve tracing by comparing the curves with those from known genuine (i.e., golden) ICs. The defects generated from the recycling process, e.g., package (MPs), bond wire (MBs), die related (MDs), and few manufacturing defects can also be detected by curve tracing. Also, a fingerprint of an IC can be generated by the traces of various combinations of pins and a decision can be made as to whether an IC is counterfeit or not by comparing its fingerprint to a genuine one.

5.3 Key Electrical Parameters Testing

Testing of the key parameters, along with functional testing for evaluating the parameters, is the most effective way of verifying the functionality of a component. These tests, which are usually conducted at room temperature (25 °C) or even higher temperatures, are generally used to test components on the manufacturing floor of assemblies in order to increase confidence that the packaged ICs are free from defects and anomalies. These tests can be useful in detecting counterfeit components, especially those re-marked to a higher grade part.

A counterfeit component may fail under these tests if any defects and anomalies are present within it. By checking the correct functionality of a component, a glut of gross defects related to leads/balls/columns, bond wires, and die related defects can easily be detected. However, these tests are perhaps the most expensive test methods available for the detection of counterfeit components when performed on complex devices. For the functional tests, a series of algorithms that exercise and test specific elements of the design are needed which requires an expensive test setup and the development of complex test programs.

If the chip has been used before (i.e., recycled counterfeit type), its DC and AC parameters may shift from their specified value mentioned on the chip's datasheet (see parametric defects EP1–EP7 in Chap. 3). After observing test results from these parametric tests, a decision can be made as to whether or not a component is counterfeit. In DC parametric tests, the parametric measurement unit (PMU) of an automatic test equipment forces the input/output voltage and current into a steady state and then measures the electrical parameters using Ohm's law. The operating

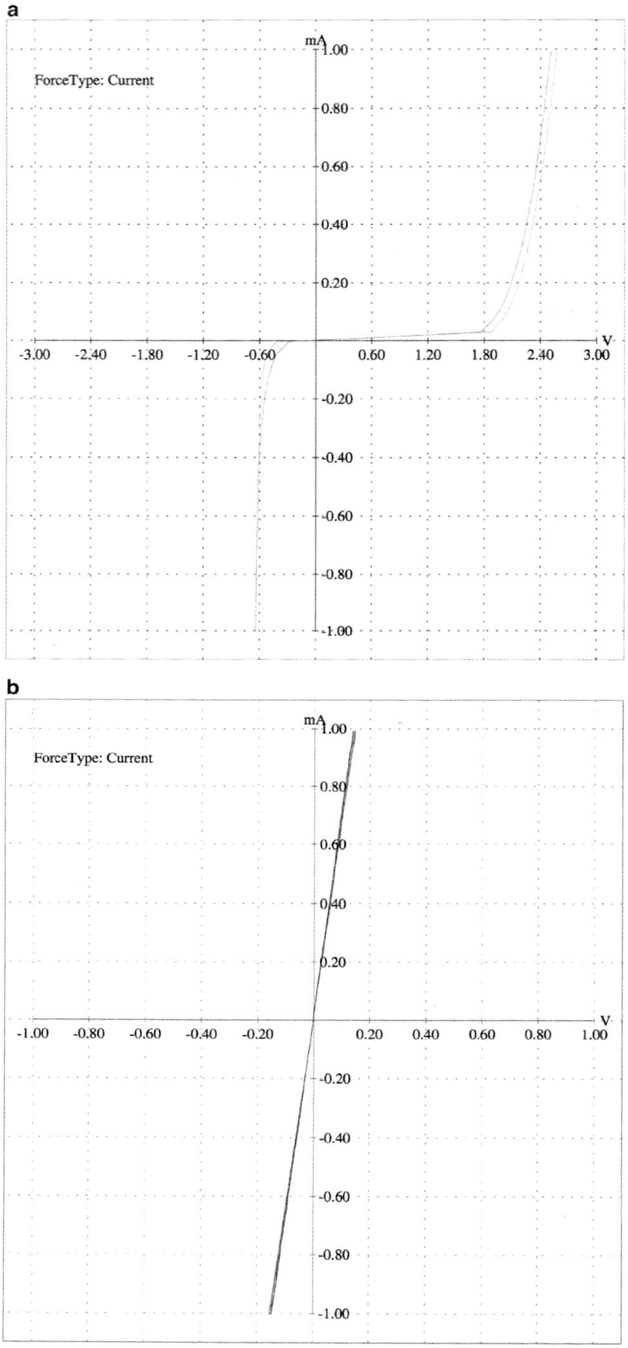

Fig. 5.4 Curve trace for Altera CPLD EPM7096QC100-15 [*Source*: Integra Technologies]. (**a**) Curve trace pass for pins 90, and 91, (**b**) curve trace fail for pins 13, 28, 40, 45, 61, 76, 88, and 97

Table 5.1 An example curve test result [Source: Integra Technologies]

Pin No	V Spot 1 (V)	I Spot 1 (mA)	V Spot 2 (V)	I Spot 2 (mA)	Result
13	−0.094	−0.605	0.086	0.592	Spot Fail
28	−0.094	−0.605	0.086	0.592	Spot Fail
40	−0.090	−0.605	0.081	0.592	Spot Fail
45	−0.094	−0.605	0.086	0.592	Spot Fail
61	−0.094	−0.605	0.086	0.592	Spot Fail
76	−0.094	−0.605	0.086	0.592	Spot Fail
88	−0.091	−0.605	0.083	0.592	Spot Fail
97	−0.094	−0.602	0.086	0.592	Spot Fail
90	−0.615	−0.605	2.350	0.592	Pass
91	−0.625	−0.605	2.410	0.592	Pass
⋮	⋮	⋮	⋮	⋮	⋮

conditions are set carefully during measurement. The DC parametric tests can be classified into the following categories: contact test, power consumption test, output short current test, output drive current test, threshold test, etc. Detailed descriptions of each test can be found in [7]. In AC parametric tests, the measurement of AC parameters (terminal impedance, timing, etc.) is performed by using AC voltages with a set of frequencies. AC parametric tests can be classified as follows: rise and fall time tests, set-up, hold and release time tests, propagation delay tests, etc. A different set of parametric tests can also be applied to memories, as in [8]. For memory, DC parametric tests include voltage bump test, leakage tests, etc., and AC parametric tests include set-up time sensitivity test, access time test, running time test, etc.

At the last phase of key electrical parameters tests, the correct functionality of a component is validated by the functional tests (see Table 5.2). Any defects that impact the functionality (from some easy-to-spot defects such as missing or broken bond wires, missing or wrong dies, etc., to more difficult-to-spot defects related to process, material, and package) can be detected. For testing of memories, read/write operations are performed to verify their functionality. MARCH tests [7, 8, 12] can be applied for counterfeit detection. Since the functions of memories are simple because of their regular structures, exhaustive functional testing is possible and is normally used during manufacturing testing [7] as shown in Table 5.2.

These tests are entirely dependent on the device type as these parameters are very different for different type of devices. For example, the parameters for SRAM will be different than a discrete component, such as a diode or a transistor. A detailed description of these tests can be found in DLA SMD 5962 standards. Table 5.2 shows the key parameters for some of the more popular devices. These parameters need to be measured and verified with the component's specification provided by the OCM. If the device fails to perform within the range specified, it should be rejected.

Table 5.2 Key electrical parameters

Device	Key parameters	Standard
Static Random Access Memory (SRAM)	1. Output HIGH/LOW Voltage (V_{OH}/V_{OL}) 2. Input/Output Leakage Current (I_{LK}/I_{OLK}) 3. V_{CC} Operating Supply Current (I_{CC}) 4. Input/Output Capacitance (C_{IN}/C_{OUT}) 5. Data Retention Current (I_{CCDR}) 6. Read/Write Timings 7. Chip Deselect to Data Retention Time (t_{DCR}) 8. Operation Recovery Time (t_R) 9. Functional algorithms: To validate proper functionality of SRAM cells (Select one or more) i. Checkerboard, checkerboard-bar ii. March iii. XY March iv. CEDES—CE deselect checkerboard, checkerboard-bar v. Equivalent algorithms depending on test labs capability	DLA SMD 5962-08219 [9]
Field Programmable Gate Arrays (FPGA)	1. High level output voltages (V_{OH1}, V_{OH2}, and V_{OH3}) 2. Low level output voltages (V_{OL1}, V_{OL2}, and V_{OL3}) 3. High/Low level input voltage (V_{IH}/V_{IL} 4. High/Low level input current ($I_{IH,IHPD}/I_{IL,ILPU}$ 5. High level Tri-state output leakage current ($I_{OZH,OZHPD}$) 6. High level output current ($I_{OZL,OZLPU}$) 7. Standby supply current (I_{CCSB}) 8. Input Capacitance (C_{IN} 9. Timings 10. Functional tests: To test the proper functionality using a serial scan test method.	DLA SMD 5962-03250 [10]
Microprocessors	1. Input voltage low (V_{ILS}, V_{ILC}) 2. Input voltage high (V_{IHS}, V_{IHC}, V_{IHR}) 3. Output voltage low (V_{OL}, V_{OLS}, V_{OLD}) 4. Output voltage high (V_{OH}, V_{OHS}) 5. Input current low (I_{IL}, I_{ILT}) 6. Input current high (I_{IH}) 7. Three-state output current high (I_{OZH}) 8. Three-state output current low (I_{OZL}, I_{OZLD}) 9. Static/Dynamic VDD supply current (I_{DD}, I_{DDOP}) 10. Input capacitance (C_{IC}, C_I) 11. Output capacitance (C_O) 12. Data bus capacitance (C_{IO}) 13. Timings 14. Functional tests	DLA SMD 5962-89519 [11]

5.4 Burn-in Testing

The reliability of a device is mainly ensured by burn-in tests [13]. In burn-in tests, the device is operated at stressed conditions to accentuate infant mortality and other unexpected failures. Such failures are often due to the latent defects, which do not necessarily expose themselves and may be skipped during manufacturing tests. Due to the electrical and thermal stresses during the usage in the field, these defects eventually expose themselves and, consequently, the devices shall fail to produce the correct functionality. During burn-in tests, the devices are operated at elevated levels of electrical (higher supply voltage) and thermal (higher temperature) stresses which accelerates the device's degradation. As a result, months to years of life time of the device are consumed in hours, allowing one to detect the presence of latent defects. Thus, by performing such tests, one can assure the reliability of a device over time as well as harsh conditions.

Burn-in test methods are described in method 1,015, MIL-STD-883 [14] for integrated circuits and methods 1,038–1,042, MIL-STD-750 [15] for other discrete components. The amount of time and the thermal conditions require to perform accelerated aging on integrated circuits are shown in Table 5.3 for the microcircuits [14]. Here, T_A represents the ambient temperature. The ambient temperature shall be 125 °C for the conditions A through E (see these test conditions in Sect. 3.1 of MIL-STD-883). The test temperatures may be increased or decreased according to Table 5.3. There are different test time requirements for class level B, S, and K components as illustrated in Table 5.3. The test times and temperatures are not necessarily fixed. The test labs that are certified and qualified to perform these tests under MIL-PRF-38535 may modify the conditions described in this table as appropriate.

The implementation of burn-in tests are very important while developing a test plan for the detection of counterfeit components as it can easily weed out the commercial grade components that have been falsely marked as military grade or space grade. It can also remove defective components or those components that were not designed to perform under stressed conditions.

5.5 Limitations and Challenges

While electrical tests have the potential to be a more efficient means of counterfeit detection than physical inspection tests, there are still many challenges associated with them. Some of the challenges overlap with those of physical tests (discussed in Chap. 4), but there are also some unique ones as well. The limitations and challenges for electrical tests are summarized as follows:

- *Process Variations*: Process variation denotes the variations in a component's parameters (e.g., length, widths, and oxide thickness of transistors) that arise from random variations occurring during fabrication. These parameters create

Table 5.3 Burn-in time-temperature regression [14]

Minimum temperature T_A(°C)	Minimum time (h)			Test condition[3]	Minimum reburn-in time (h)
	Class level S[1]	Class level B[2]	Class level S hybrids (Class K)		
100	—	352	700	Hybrids only	24
105	—	300	600	"	24
110	—	260	520	"	24
115	—	220	440	"	24
120	—	190	380	"	24
125	240	160	320	A–E	24
130	208	138	—	"	21
135	180	120	—	"	18
140	160	105	—	"	16
145	140	92	—	"	14
150	120	80	—	"	12
175	—	48	—	F	12
200	—	28	—	"	12
225	—	16	—	"	12
250	—	12	—	"	12

[1] High reliability military applications (Class level B).
[2] Space applications (Class level S).
[3] Test Conditions defined in Section 3.1 MIL-STD-883 [14]

differences in the component's performance from the nominal (designed) performance. With rapid semiconductor scaling, the electrical parameters (see Table 5.2) of modern ICs may vary significantly. Thus, it is becoming more difficult to determine whether the variations in the parameters of a component are due to the counterfeiting (e.g., recycled, remarked, cloned, etc. counterfeit types) or unavoidable process variations. One can perform a statistical analysis based on the data observed from the parametric tests to determine the confidence level that a part is counterfeit with or without authentic or golden ICs. The efficiency of such analysis must be proven on a large number of golden and counterfeit ICs.
- *Test Time and Cost*: Burn-in tests are useful in detecting infant mortality failures of components as we described above. However, because of excessive test time (tens of hours), requirement of a high-speed tester in order to apply functional test patterns to complex chips, etc. electrical tests can also be extremely expensive. Thus, electrical tests of complex components might only attractive for critical and high-risk applications. Finally, as discussed above, ATEs require highly specialized algorithms and test programs. The sheer number of component types (digital ICs, analog ICs, mixed ICs, discrete components, etc.) also makes it challenging, if not impossible, to create an all-in-one setup and/or programs to detect each type.
- *Lack of Part Specifications*: In many instances, one cannot get a hold of the complete set of test vectors to test an obsolete part as the original component manufacturer (OCM) may no longer exist or the information needed may no longer be available in archived records at the OCM. Test program generation for obsolete and active parts without this knowledge is extremely difficult, if not impossible.
- *Counterfeit Types*: Not all counterfeit types are adequately covered by electrical tests. For example, overproduced, cloned, and tampered ICs will avoid detection as long as their electrical parameters and performance remain within the component specification.

Preliminary work on overcoming many of the above challenges is contained in latter chapters of this book. These include assessment and optimization of tests (Chap. 6), advanced electrical tests (Chap. 8), and design-for-anti-counterfeit (Chaps. 9–12).

5.6 Summary

In this chapter, we have discussed all the popular electrical tests currently recommended by various standards. The main advantage of electrical tests is their nondestructive nature. That being said, they have many limitations. First, they are completely dependent on the part type and require different test setups for different components. This makes the electrical tests extremely expensive as one must incur the non-recurring expenses (NRE) due to the variety of test equipment and test fixtures needed. As a result, electrical tests are often performed as one of the last steps of the detection process (when the volume of components under authentication is already small). Second, they often require direct comparison to the OCM spec sheet which is not always available. Finally, process variation can mask the parameter variations seen within all counterfeit components (recycled, remarked, out-of-spec/defective, overproduced, cloned, tampered).

The challenges and limitations presented in this chapter along with those from Chap. 4 strongly suggest that there is a need to quantify all tests in terms of time, cost, coverage of defects, and coverage of counterfeit types. Quantitative metrics could be used to select the "best" combination of physical and electrical tests in order to keep test time and cost low while also providing adequate coverage of all counterfeit defects and types. The first assessment framework for counterfeit detection along these lines shall be discussed in the next chapter.

References

1. U. Guin, M. Tehranipoor, D. DiMase, M. Megrdichian, Counterfeit IC detection and challenges ahead. ACM/SIGDA E **43**(3), 38–43 (2013)
2. U. Guin, D. DiMase, M. Tehranipoor, A comprehensive framework for counterfeit defect coverage analysis and detection assessment. J. Electron. Test. **30**(1), 25–40 (2014)
3. U. Guin, D. DiMase, M. Tehranipoor, Counterfeit integrated circuits: detection, avoidance, and the challenges ahead. J. Electron. Test. **30**(1), 9–23 (2014)
4. U. Guin, K. Huang, D. DiMase, J. Carulli, M. Tehranipoor, Y. Makris, Counterfeit integrated circuits: a rising threat in the global semiconductor supply chain. Proc. IEEE **102**(8), 1207–1228 (2014)
5. U. Guin, M. Tehranipoor, On selection of counterfeit IC detection methods, in *IEEE North Atlantic Test Workshop (NATW)*, (2013)
6. A. Grochowski, D. Bhattacharya, T. Viswanathan, K. Laker, Integrated circuit testing for quality assurance in manufacturing: history, current status, and future trends. IEEE Trans. Circuits Syst. II **44**(8), 610–633 (1997)
7. M. Bushnell, V. Agrawal, *Essentials of Electronic Testing for Digital, Memory, and Mixed-Signal VLSI Circuits* (Springer, New York, 2000)
8. P. Mazumder, K. Chakraborty, *Testing and Testable Design of High-Density Random-Access Memories* (Springer, New York, 1996)
9. DLA, MICROCIRCUIT, MEMORY, DIGITAL, CMOS, 2M x 8-BIT (16M), STATIC RANDOM ACCESS MEMORY (SRAM), (3.3 V), MONOLITHIC SILICON, http://www.landandmaritime.dla.mil/downloads/milspec/smd/08219.pdf (2009). DRAWING APPROVAL DATE: 27 Feb 2009
10. DLA, SMD 5962-03250: MICROCIRCUIT, MEMORY, DIGITAL, CMOS, FIELD PROGRAMMABLE GATE ARRAY, 40,000 GATES WITH 18 K OF INDEPENDENT SRAM, MONOLITHIC SILICON, http://www.landandmaritime.dla.mil/Downloads/MilSpec/Smd/03250.pdf (2004). DRAWING APPROVAL DATE: 04 May 2004
11. DLA, SMD 5962-89519: MICROCIRCUIT, DIGITAL, CMOS, 16-BIT MICROPROCESSOR, MIL-STD-1750 INSTRUCTION SET ARCHITECTURE, MONOLITHIC SILICON, http://www.landandmaritime.dla.mil/downloads/milspec/smd/89519.pdf (1989). DRAWING APPROVAL DATE: 15 Feb 1989
12. D. Suk, S. Reddy, A march test for functional faults in semiconductor random access memories. IEEE Trans. Comput. **C-30**(12), 982–985 (1981)

13. F. Jensen, N.E. Petersen, *Burn-in: An Engineering Approach to the Design and Analysis of Burn-In Procedures* (Wiley, Chichester, 1982)
14. Departnent of Defense, Test Method Standard: Microcircuits, (2010), http://www.landandmaritime.dla.mil/Downloads/MilSpec/Docs/MIL-STD-883/std883.pdf
15. Departnent of Defense, Test Method Standard: Test Methods for Semiconductor Devices, (2012), http://www.landandmaritime.dla.mil/Downloads/MilSpec/Docs/MIL-STD-750/std750.pdf

Chapter 6
Counterfeit Test Coverage: An Assessment of Current Counterfeit Detection Methods

The detection of counterfeit integrated circuits has become a major challenge largely due to the deficiencies in today's testing mechanisms [1–4]. The detection of such components is still in its infancy, and there are major challenges that must be overcome in order for effective counterfeit detection methods to be deployed. Counterfeiting is an evolving problem with counterfeiters acquiring increasing amounts of experience with each passing day. Hence, it is imperative that we make every effort to stay ahead of them in order to prevent the widespread infiltration of counterfeit parts into our critical infrastructures. By detecting counterfeit parts efficiently, we can also enhance the public's confidence in the security of systems that surround them. In order to achieve this goal, we must be able to continuously monitor counterfeiting activity and assess counterfeit detection methods in order to evaluate their effectiveness in detecting counterfeit components. We also need to develop a common platform to evaluate the efficacy of a set of test methods.

In this chapter, we will discuss the metrics for evaluating counterfeit test methods:

i. **Counterfeit Defect Coverage (CDC)** represents the confidence of detecting defects by a set of test methods.
ii. **Counterfeit Type Coverage (CTC)** represents the confidence of detecting specific counterfeit types by a set of test methods.
iii. **Under-Covered Defects (UCDs)** represents the defects that are partially detected by a given set of tests.
iv. **Not-Covered Defects (NCDs)** represents the defects that are not detected by a given set of tests.

We will also present a comprehensive framework for (i) assessing a set of test methods to evaluate their effectiveness based on these newly developed metrics. We call this as "static assessment"; (ii) selecting a set of test methods to maximize the test coverage considering test cost and time budget; and (iii) deciding on the best

M. Tehranipoor et al., *Counterfeit Integrated Circuits*,
DOI 10.1007/978-3-319-11824-6_6

set of test methods for achieving maximum test coverage. We call the combination of (ii) and (iii) as "dynamic assessment". These metrics and the assessment of test methods were initially introduced in [5, 6].

6.1 Disparity in Capabilities and Expertise Among Test Labs

Assessing the capabilities of different test labs is indeed an important requirement. Honeywell performed round robin testing in 2012 and 2013 to certify test labs and evaluate their capabilities [7, 8]. In 2012, they gave five samples of one counterfeit part (National Semiconductor DAC1230LCJ) and one authentic part (Tundra CA91L860B-50CE) to 12 test labs (A to M in Table 6.1). The labs were encouraged to process the parts as per their standard flow with no special processing. Columns 2 and 3 in Table 6.1 show the results of identifying those components as counterfeit or authentic. The × and √ marks represent the incorrect and correct identification of a part by a test lab whereas *NA* denotes that a test lab did not participate in this evaluation. Test Labs A and K failed to identify the authentic component, whereas Test Lab E missed the counterfeit component. In 2013, Honeywell performed this assessment again with six test labs providing five samples of two counterfeit parts (Intel TB28F400B5T80 flash memory, and

Table 6.1 Test lab comparison

Test lab	National semiconductor DAC1230LCJ (Counterfeit, 2012)	Tundra CA91L860B-50CE (Authentic, 2012)	Intel TB28F400B5T80 (Counterfeit, 2013)	TDK C5750Y5V1H226Z (Counterfeit, 2013)
A	√	×	√	√
B	√	√	NA	NA
C	√	√	√	√
D	√	√	√	√
E	×	√	NA	NA
F	√	√	NA	NA
G	No conclusion stated	No conclusion stated	NA	NA
H	√	√	NA	NA
I	√	√	NA	NA
J	√	√	√	√
K	√	×	NA	NA
L	NA	NA	√	√
M	√	√	√	√

TDK C5750Y5V1H226Z capacitance). All the test labs correctly identified these counterfeit parts. However, they missed several significant potential indicators of counterfeiting.

The following were the conclusions drawn from the above test lab comparison [7]: (i) the test labs showed improved performance in identifying counterfeit parts as they gained greater experience and exposure to different counterfeit parts, and (ii) these labs were able to accurately detect some easy-to-detect counterfeit defects related to blacktopping, dimension and color variations, and solder issues, but they had more difficulty with hard-to-detect defects related to lead finish, dimple depth, improper materials, and electrical parameters. For some labs, defect identification was as low as 32 %. Thus, there is a need to assess test labs' capability in terms of quantitative measures, which will finally lead to the development of test metrics.

6.2 Terminologies

The purpose of assessing test methods is to establish the effectiveness of the testing currently being performed to detect counterfeit components. To make it easy for the reader to understand the assessment process, we will first describe several key elements (tier level, target confidence, confidence level matrix, defect frequency, decision index, and defects mapping matrix) below. All these elements will also represent inputs to our proposed assessment framework in later sections.

6.2.1 Tier Level

Tier level (TL) was introduced in AS6171 [9] as a means of assessing the risk associated with the use of a part while also determining the recommended level of testing that should be performed. While assessing risk, three main factors was taken into consideration: (i) the final product in which a part shall be used, (ii) the functionality of a part within a product, and (iii) quality attributes associated with the supplier that sells or distributes parts to various entities in the electronics supply chain. Thus, TL is calculated from product risk (RP), component risk (RC), and supplier risk (RS). The risk associated with a component's application is characterized by both RP and RC. The probability of receiving a counterfeit component from a supplier is addressed by RS. Column 3 of Table 6.2 shows the risk scores for different tier levels. The detailed calculation of calculating the scores can be found in the risk assessment section of AS6171 [9]. Depending on this risk category, a different set of test methods are recommended for different tier levels in AS6171 [9]. It is extremely important for user/requester to know the tier level they belong, before implementing a test plan for the screening of counterfeit parts.

Table 6.2 Risk scores for different tier levels [9]

Tier levels (TL)	Risk category	Score range	Target confidence (TC)
4	Critical	> 170	0.95
3	High	151–170	0.8
2	Moderate	111–150	0.65
1	Low	71–110	0.5
0	Very low	0–70	0.35

6.2.2 *Target Confidence*

The target confidence (TC) for each defect is the level of confidence achieved after performing a set of tests. The value of TC for each tier level is shown in Column 4 of Table 6.2. TC increases from very low to critical tier applications. We need to have higher levels of test confidence for each defect in order for higher tier levels to increase the overall level of test confidence. Based on this confidence, we will develop under-covered defects (Sect. 6.3.4) and guide dynamic test assessment (Sect. 6.4.2).

6.2.3 *Test Methods*

The tests methods under assessment are outlined in Chaps. 4 and 5 (see Fig. 4.1). Each test method has an associated "cost" and "time", defined as the cost and time involved in testing one batch of components. For method i, we shall denote the cost and time by C_i and T_i respectively.

6.2.4 *Counterfeit Defects*

Counterfeit defects are defined as the defects and anomalies seen in electronic components. These defects were thoroughly described in Chap. 3 (see Fig. 3.1). Test methods are assessed based on their ability to detect one or more defects, and test confidence increases as the number of detected defects increases. In our notation, D_j denotes the jth defect from the defects taxonomy.

6.2.5 *Confidence Level Matrix*

The confidence level matrix (CL) represents the capability of test methods to detect counterfeit defects. When a test is performed, it detects some of the counterfeit defects. However, it does not necessarily follow that the same test will detect the same defects in different counterfeit components. Generally a confidence (or probability of detection) is involved in this detection process. In this CL matrix, each entry represents a certain level of confidence for detecting a defect by a given test method.

$CL = [x_{ij}]$, where x_{ij} is the probability of detecting a defect j by a method i. Here, the rows and columns of CL are denoted as the methods and the defects, respectively.

If two or more methods detect the same defect, then the resultant confidence level (x_{Rj}) will be increased and is given by the following equation,

$$x_{Rj} = 1 - \prod_{i=1}^{m_s} \left(1 - x_{ij}\right) \quad \text{for defect } j \tag{6.1}$$

where m_s represents the number of tests in the recommended test set.

6.2.6 Defect Frequency

Defect frequency (DF) is defined as how frequently the defect is visible in a counterfeit component. This is one of the key parameters for evaluating test coverage, as the detection of high frequency defects has more of an impact than the detection of low frequency defects.

6.2.7 Decision Index

The decision index (DI) is defined as the probability that a counterfeit type contains one or more known counterfeit defects. It can also be interpreted as the probability of identifying a component belonging to a counterfeit type after targeting all defects. It is not necessarily true that every occurrence of a counterfeit type will contain a one or more defects. For example, DI may approach zero for certain counterfeit types—such as overproduced and cloned counterfeit types—due to the rare occurrence of defects in these types. Table 6.3 shows the DI values for different counterfeit types.

6.2.8 Defect Mapping Matrix

The defect mapping (DM) matrix represents the presence of a defect in a counterfeit type. It is not necessary for a defect to be visible in all the counterfeit types simultaneously. For example, the defect *Invalid Lot/Date/Country Code* may not

Table 6.3 Decision index for each counterfeit type

Counterfeit type	Decision index
Recycled	0.98
Remarked	0.90
Overproduced	0.03
Out-of-spec/defective	0.98
Cloned	0.10
Forged documentation	0.70
Tampered	0.98

Table 6.4 Terminologies used in our proposed method selection algorithm

Terminology	Notation
Test methods	$M = [M_1\ M_2\ \dots\ M_m]$, where m is the number of test methods
Test cost	$C = [C_1\ C_2\ \dots\ C_m]$
Test time	$T = [T_1\ T_2\ \dots\ T_m]$
Counterfeit defects	$D = [D_1\ D_2\ \dots\ D_n]$, where n is the number of defects
Tier level	$TL = [L_1\ L_2\ \dots\ L_5]$, L_1: Critical, L_2: High, L_3: Moderate, L_4: Low, L_5: Very low
Target confidence	$TC = [TC_1\ TC_2\ \dots\ TC_5]$, TC_1: Critical, TC_2: High, TC_3: Moderate, TC_4: Low, TC_5: Very low
Confidence level matrix	$CL = [x_{ij}] = \begin{bmatrix} x_{11} & x_{12} & \dots & x_{1n} \\ x_{21} & x_{22} & \dots & x_{2n} \\ \vdots & \vdots & \vdots & \vdots \\ x_{m1} & x_{m2} & \dots & x_{mn} \end{bmatrix}$ where, $x_{ij} = Pr$ (Detecting a defect j by a method i). Here, the rows and columns of CL are denoted as the methods and the defects
Defect frequency	$DF = [DF_1\ DF_2\ \dots\ DF_n]$
Decision index	$DI = [DI_1\ DI_2\ \dots\ DI_7]$
Defect mapping matrix	$DM = [w_{ij}] = \begin{bmatrix} w_{11} & w_{12} & \dots & w_{17} \\ w_{21} & w_{22} & \dots & w_{27} \\ \vdots & \vdots & \vdots & \vdots \\ w_{n1} & w_{n2} & \dots & w_{n7} \end{bmatrix}$ where, $w_{ij} \in \{0, 1\}$ ={Not Present, Present}, and rows and columns represent defects and counterfeit types respectively

be present in overproduced or cloned types. Here, each entry of DM equals 1 if the defect may be present for a counterfeit type and 0 if the defect is never present.

In summary, Table 6.4 shows all the notation described above that will be used in the assessment framework.

6.2.9 Challenges Associated with Input Acquisition

The detection of counterfeit parts is still in its infancy and the data for most of the key elements discussed above does not exist anywhere today. In addition, based on the evolution of counterfeiters and counterfeiting practices, much of this data is dynamic. For example, the appearance of different counterfeit types, the defect frequencies, etc. will change over time. The number of test methods may also increase over in the future as well as their detection capabilities. In Chap. 7, we propose some methods that could lead to automation in collecting this data. In the near future, some of this data may also become available as the two reporting agencies ERAI [10] and GIDEP [11] are capturing the incidence of counterfeit parts worldwide. It is important to mention that the entire assessment framework (metrics and algorithms) we describe below is flexible enough to handle any changes to the input data and/or its source.

6.3 Test Metrics

To evaluate the effectiveness of these test methods, it is of the utmost importance to develop test metrics that represent coverage for detecting counterfeit defects. These metrics are described in detail below.

6.3.1 Counterfeit Defect Coverage (CDC)

Counterfeit defect coverage (CDC) is defined as the ratio of all probable detectable defects by a set of (single) test methods (method) to the total number of known counterfeit defects. It provides a cumulative confidence of identifying a component as counterfeit by a sequence of test methods. Intuitively, CDC can be described as follows:

$$CDC = \frac{\text{Probable Detectable Defects}}{\text{Total Defects}} \times 100\,\% \tag{6.2}$$

A level of confidence is involved when a test method detects a counterfeit defect, and this is captured in the CL matrix. When a defect is detected by multiple test methods, the confidence of identifying it increases. The maximum value of this confidence is bounded by "1", which signifies this particular defect will surely be detected by these test methods. Now the total confidence of finding a part as counterfeit, CDC, becomes the ratio of the cumulative sum of the resultant confidence of all the defects to the total number of defects. Thus, in our notation, it can be expressed as:

$$CDC = \frac{\sum_{j=1}^{n}(x_{Rj})}{n} \times 100\,\% \tag{6.3}$$

where x_{Rj} denotes the resultant confidence for defect j and n denotes total number of defects.

Equation (6.3) (shown above) treats all the defects as equally likely in the component supply chain. However, some defects are more frequent than others, and we need to incorporate defect frequency in the calculation of CDC. Therefore, the modified equation of CDC becomes,

$$CDC = \frac{\sum_{j=1}^{n}(x_{Rj} \times DF_j)}{\sum_{j=1}^{n} DF_j} \times 100\,\% \tag{6.4}$$

where DF_j represents the defect frequency for defect j.

6.3.2 Counterfeit Type Coverage (CTC)

Defects are not equally visible in all counterfeit types. Some defects may be present in some particular counterfeit types, but, not in other types, which is captured in defects mapping (DM) matrix, and for some counterfeit types, the probability of finding any defects is extremely small, which is captured in decision index (DI). For example, overproduced parts may be as good as new authentic parts and be free from any counterfeit defects. As such, the detection of defects does not necessarily provide the correct test coverage, CDC, which was introduced in Sect. 6.3.1. We will now introduce, counterfeit type coverage (CTC) to represent the test coverage for individual counterfeit types by a set of test methods.

CTC is a measure to detect a counterfeit type (recycled, remarked, etc. as described in Chap. 2, Sect. 2.2) given the test methods performed. CTC can be expressed as:

$$CTC = DI \times \frac{\text{Probable Detectable Defects for a Counterfeit Type}}{\text{Total Defects for a Counterfeit Type}} \times 100\,\% \tag{6.5}$$

where, DI represents the decision index.

CTC can also be described as the total confidence of finding a part that belongs to a particular counterfeit type. Taking CTC for a counterfeit type k becomes the ratio of the cumulative sum of the resultant confidence of all the defects detected by the test methods to the total number of defects belonging to that counterfeit type:

$$\mathrm{CTC}_k = \mathrm{DI}_k \times \frac{\sum_{j=1}^{n}(x_{Rj} \times w_{jk})}{\sum_{j=1}^{n}(w_{jk})} \times 100\,\% \tag{6.6}$$

where,

CTC_k : CTC for counterfeit type k;
DI_k : DI for counterfeit type k;
x_{Rj} : Resultant confidence for defect j;
w_{jk} : The presence of defect j in counterfeit type k ($\in \{0, 1\}$).

Equation (6.6) (shown above) treats all the defects as equally likely in the component supply chain. However, some defects are more frequent than others, so we need to incorporate defect frequency in the calculation of CTC. The modified equation of CTC becomes

$$\mathrm{CTC}_k = \mathrm{DI}_k \times \frac{\sum_{j=1}^{n}(x_{Rj} \times DF_j \times w_{jk})}{\sum_{j=1}^{n}(DF_j \times w_{jk})} \times 100\,\% \tag{6.7}$$

where, DF_j: Defect frequency of defect j.

6.3.3 Not-Covered Defects (NCDs)

A set of test methods will not necessarily detect a particular counterfeit defect. A defect is called as a not-covered defect (NCD) when a set of test methods cannot detect it. A counterfeit defect j will be a NCD if

$$x_{Rj} = 0 \tag{6.8}$$

where x_{Rj} is the resultant confidence for defect j and is given by Eq. (6.1).

6.3.4 Under-Covered Defects (UCDs)

A defect is called an under-covered defect (UCD) when a set of test methods cannot provide a desired confidence level. Defects belong to this category when the resultant confidence level for detecting a defect is less than the target defect confidence level. A defect j will be a UCD if

$$x_{Rj} < TC \tag{6.9}$$

where x_{Rj} is the resultant confidence for defect j and is given by Eq. (6.1), and TC is the target confidence for each defect.

6.4 Assessment Framework

Different sequences of test methods have been developed by organizations for the detection of counterfeit parts. The assessment framework evaluates the effectiveness of a sequence of test methods used to screen for counterfeit parts. This framework works in two different modes. In the static assessment, it performs the assessment of a sequence of tests under evaluation. The output of this mode produces the test metrics (CDC, CTC, NCD, and UCD). In the dynamic assessment, the framework receives all the current available test methods as input and recommends (i) the best set of tests and (ii) an optimum set of tests that provides maximum coverage within a certain test time and cost budget. Then the assessment is done on the basis of the same test metrics.

6.4.1 *Static Assessment*

As it is necessary to evaluate the capability of the test labs, the static assessment provides the test labs with an evaluation of the effectiveness of a specified test plan consisting of a sequence of tests. The term "static" suggests that, in this kind of assessment, the test methods put into this framework do not change, and the assessment is performed on this whole set of test methods.

6.4.1.1 Assessment of Test Methods

Algorithm 1 shows the flow of this assessment framework. The user specified test methods are provided to this framework as inputs. It selects the target confidence from the user specified risk category (tier level breakpoints). It then reads the confidence level matrix (CL), decision index (DI), and defects mapping matrix (DM) from a secured database. The function CALCULATE() in line 3, calculates the resultant confidence level for all the defects. The CALCULATE() functions in lines 5–8, calculate CDC, CTC, $NCDs$ and $UCDs$.

Algorithm 1 Static Assessment

1: Inputs: User specified test methods (M^S), confidence level matrix (CL), decision index (DI), and defects mapping matrix (DM)
2: **for** (all defect index j from 0 to n in D) **do**
3: Calculate x_{Rj}, $x_{Rj} \leftarrow$ CALCULATE (X, M^S)
4: **end for**
5: Calculate counterfeit defect coverage, $CDC \leftarrow$ CALCULATE (x_R, DF)
6: Calculate counterfeit type coverage, $CTC \leftarrow$ CALCULATE (x_R, DF, DI, DM)
7: Calculate not-covered defects, $NCDs \leftarrow$ CALCULATE (x_R)
8: Calculate under-covered defects, $UCDs \leftarrow$ CALCULATE (x_R, TC)
9: Report CDC, CTC, $NCDs$ and $UCDs$

6.4.1.2 Example

In this section, we will discuss a short example with synthetic data. Let us assume that we want to assess five test methods for critical tier level (tier 4, described in Table 6.2). We also assume that there are five test methods ({M1, M2, M3, M4, M5}) present for counterfeit detection and five counterfeit defects ({D1, D2, D3, D4, D5}) present in the counterfeit parts with a given confidence level (CL) matrix and defect frequency vector of

$$CL = \begin{array}{c} \\ M1 \\ M2 \\ M3 \\ M4 \\ M5 \end{array} \begin{array}{c} \begin{array}{ccccc} D1 & D2 & D3 & D4 & D5 \end{array} \\ \begin{bmatrix} 0.9 & 0.5 & 0.0 & 0.0 & 0.0 \\ 0.0 & 0.0 & 0.9 & 0.0 & 0.5 \\ 0.0 & 0.9 & 0.0 & 0.0 & 0.0 \\ 0.9 & 0.0 & 0.0 & 0.0 & 0.0 \\ 0.0 & 0.0 & 0.9 & 0.0 & 0.0 \end{bmatrix} \end{array}, \quad \text{and} \quad DF = \begin{array}{c} D1 \\ D2 \\ D3 \\ D4 \\ D5 \end{array} \begin{bmatrix} 1 \\ 1 \\ 1 \\ 1 \\ 1 \end{bmatrix}$$

This matrix can be interpreted as follows: Each row represents a test method (e.g., the first row represents the test method M1, the second row represents the test method M2, and so on). Each column represents a defect (e.g., the first column represents defect 1, the second column represents defect 2, and so on). Each entry denotes the confidence of detecting a defect using a test method. This means that test M1 has a 0.9 probability of detecting defect D1, a 0.5 probability of detecting defect D2, and a 0 probability of detecting defects D3, D4, and D5.

We also assume that there are three counterfeit types ({x,y,z}) with defect mapping (DM) matrix and decision index (DI) vectors of

$$DM = \begin{array}{c} \\ D1 \\ D2 \\ D3 \\ D4 \\ D5 \end{array} \begin{array}{c} \begin{array}{ccc} x & y & z \end{array} \\ \begin{bmatrix} 1 & 0 & 1 \\ 0 & 0 & 1 \\ 1 & 1 & 0 \\ 1 & 1 & 0 \\ 1 & 1 & 1 \end{bmatrix} \end{array}, \quad \text{and} \quad DI = \begin{array}{c} x \\ y \\ z \end{array} \begin{bmatrix} 0.9 \\ 0.5 \\ 0.1 \end{bmatrix}$$

Table 6.5 summarizes the assessment process. The CDC is 68.8 %, whereas the CTC for counterfeits x, y, and z is 55.8, 24.8, and 8.1 %, respectively. The lower value of CTC for counterfeits y and z points to the fact that the defects related to those counterfeit types are not visible as we can see from the DI vector that the probability of finding any counterfeit defect is 0.5 and 0.1, respectively.

6.4.1.3 Results

The University of Connecticut's CHASE Center has implemented this framework to assess the effectiveness of the test methods being recommended by different test labs

Table 6.5 Sample example for the static assessment

Step	Name	Description
Line 1	Read inputs	Read M^S, TL, CL, DI, and DM
Line 2–4	Compute resultant confidence (x_R) using Eq. (6.1)	$x_{RD1} = 1 - \{(1-0.9)(1-0)(1-0)(1-0.9)(1-0)\} = 0.99$ $x_{RD2} = 1 - \{(1-0.5)(1-0)(1-0.9)(1-0)(1-0)\} = 0.95$ $x_{RD3} = 1 - \{(1-0)(1-0.9)(1-0)(1-0)(1-0.9)\} = 0.99$ $x_{RD4} = 1-\{(1-0)(1-0)(1-0)(1-0)(1-0)\} = 0.00$ $x_{RD5} = 1 - \{(1-0)(1-0.5)(1-0)(1-0)(1-0)\} = 0.50$
Line 5	Compute CDC using Eq. (6.4)	$CDC = 100 * \frac{1*0.99+1*0.95+1*0.99+1*0.00+1*0.50}{1+1+1+1+1+1}\,\% = 68.6\,\%$
Line 6	Compute CTC using Eq. (6.7)	$CTC_x = 0.9 * \frac{1*0.99+0*0.95+1*0.99+1*0.00+1*0.5}{1+0+1+1+1} \times 100\,\% = 55.8\,\%$ $CTC_y = 0.5 * \frac{0*0.99+0*0.95+1*0.99+1*0.00+1*0.5}{0+0+1+1+1} \times 100\,\% = 24.8\,\%$ $CTC_z = 0.5 * \frac{1*0.99+1*0.95+0*0.99+0*0.00+1*0.5}{1+1+0+0+1} \times 100\,\% = 8.1\,\%$
Line 7	Compute $NCDs$ using Eq. (6.8)	NCD: Defect D4 as $x_{Rd} = 0$
Line 8	Compute $UCDs$ using Eq. (6.9)	UCD: Defect D5 as $x_{Re} < TC$ (0.50 < 0.95))

and standards. In this section, we are going to present a sequence of test methods for low risk category to test microcircuits, recommended by SMEs.

As discussed in Sect. 6.2.9, most of the data required by the assessment framework does not exist. In order to evaluate the proposed approaches, we have compiled information from the subject matter experts (SMEs) and test labs participating in the G-19A group [12]. This information includes confidence matrix (CL), decision index (DI), defect frequency (DF), etc. We shall not show this data here as this information could cause harm in the hands of counterfeiters. Since the test time and cost vary across different test labs, we use the average over all test labs as input to our framework.

Table 6.6 shows the assessment of the test methods for the low risk category. Column 3 represents the CDC. General EVI alone contributes a coverage of 35.4 %. General EVI and detailed EVI contribute a combined coverage of 47.6 %. The combined coverage of the total 11 test methods gives a final CDC of 82.1 %. This signifies the fact that we are 82.1 % confident of finding a part as counterfeit after performing these test methods. There are three defects that cannot be detected by these tests and become $NCDs$. Except for the $NCDs$, all the defects are covered as the resultant confidence levels for these test methods become larger than the target confidence. Thus, there are no $UCDs$.

Table 6.7 shows the $CTCs$ for all the counterfeit types. As we explained before, detecting defects can help us identify a component as counterfeit. However, this

Table 6.6 Assessment of test methods for low risk category (CDC, NCDs, and UCDs)

#	Test methods	CDC (%)	NDCs	UCDs
1	General EVI	35.4		
2	Detailed EVI	47.6		
3	Testing for remarking (EVI)	48.1		
4	Testing for resurfacing (EVI)	48.3		
5	Lead finish analysis (XRF)	48.5		
6	Lead finish thickness (XRF)	48.5		
7	Material composition (XRF)	51.4		
8	Internal inspection	65.4		
9	Radiological inspection	71.5		
10	Acoustic microscopy	71.6		
11	DC test at ambient temperature	82.1		
	Test plan for low risk category	82.1	3	0

Table 6.7 Assessment of test methods for low risk category (CTC)

#	Counterfeit type	CTC
1	Recycled	82.8
2	Remarked	84.5
3	Overproduced	1.6
4	Out-of-spec./defective	53.3
5	Cloned	6.3
6	Forged documentation	68.9
7	Tampered	NA

cannot provide the necessary confidence that the counterfeit component belongs to a particular type. CTC gives what is desired for finding a counterfeit type. The $CTCs$ for recycled and remarked types are close to CDC, as the probability of finding any counterfeit defects is close to 1 (i.e., 0.98 and 0.9 for recycled and remarked types indicated in the DI vector in Table 6.3). However, in overproduced and cloned types, the $CTCs$ are quite small and are 1.6, and 6.3 %, respectively. The probability of finding counterfeit defects in these counterfeit types is extremely small. This signifies that we need a different set of measures (design-for-anti-counterfeit, DFAC) to detect these counterfeit types. We will introduce different DFAC measures in Chaps. 9, 11, 12, and 10. We have not assessed tampered types, as they provide a different set of challenges for understanding their defects and anomalies. We use the term "not applicable" (NA) in the CTC field for tampered types in Table 6.7.

6.4.2 Dynamic Assessment

We need to identify a set of test methods targeting critical risk applications in order to maximize the test confidence (i.e., our ability to detect counterfeit parts). For critical applications (aerospace, military, transportation, etc.), there should be as little margin for error as possible. At the same time, when the risk category level is low, the user does not need to engage in exhaustive testing. In this case, test time and cost are more important, and we need to find the best set of tests that can give the maximum coverage under these test time and cost constraints. In the following, we will first present the method selection technique and then we will assess those selected test methods.

6.4.2.1 Selection of Test Methods

The objective of the method selection algorithm is to find an optimum set of methods to maximize counterfeit defect coverage while also allowing for a consideration of the test time, cost, and risk category constraints for certain applications. A counterfeit defect can be detected by multiple methods with different levels of confidence. Thus, the problem becomes that of selecting the most suitable methods for achieving the highest CDC possible given the presence of practical constraints.

The problem can be formulated as follows:

Select a set of methods $M^S \subset M$ *to Maximize CDC*

Subjected to:

$$x_{Rj} \geq TC, \ \forall\ j \in \{1:n\} \qquad \textit{for critical applications}$$

or

$$\begin{cases} x_{Rj} \geq TC, \ \forall\ j \in \{1:n\} \\ M_1C_1 + M_2C_2 + \ldots + M_mC_m \leq C_{user} \\ M_1T_1 + M_2T_2 + \ldots + M_mT_m \leq T_{user} \end{cases} \qquad \textit{for non-critical applications}$$

where,

x_{Rj}: Resultant confidence for defect j;
TC: Target confidence;
M_i: Test method i, $M_i \in \{0, 1\} = \{\text{Not Selected, Selected}\}$;
C_i: Test cost for test method i;
T_i: Test time for test method i;
m: Number of test methods;
n: Number of defects;
C_{user}: User specified total test cost;
T_{user}: User specified total test time;

Algorithm 2 describes the selection of the test methods. It starts by initializing the recommended test set to null. It then gets the defect frequency (DF) and the target confidence level (TC) for each defect. Next, it prioritizes the defects by sorting them according to DF, as we want to capture high-frequency defects first to achieve a higher CDC.

Algorithm 2 Selection of Test Methods

1: Initialize selected methods, $M^S \leftarrow \{\phi\}$.
2: Specify cost limit set by the user c_{user} except for critical risk applications
3: Specify test time limit set by the user t_{user} except for critical risk applications
4: Specify risk category, $TC \leftarrow$ user specified tier level.
5: Get confidence level matrix (CL).
6: Get defect frequency (DF).
7: Sort defects according to defect frequency, $D \leftarrow$ SORT (DF)
8: **if** $(TL ==$ critical$)$ **then**
9:     **for** (all defect index j from 0 to n in D) **do**
10:         Sort methods according to x_{ij}, $M' \leftarrow$ SORT (M, CL)
11:         Calculate x_{Rj}, $x_{Rj} \leftarrow$ CALCULATE (CL, M')
12:         **for** (all method index i from 0 to m in M')) **do**
13:             SELECTMETHODS (CL, M', x_{Rj}, TC)
14:         **end for**
15:     **end for**
16: **else**
17:     **for** (all defect index j from 0 to n in D) **do**
18:         Sort methods according to test time and cost, $M' \leftarrow$ SORT (M, T, C)
19:         Calculate x_{Rj}, $x_{Rj} \leftarrow$ CALCULATE (CL, M')
20:         **for** (all method index i from 0 to m in M')) **do**
21:             SELECTMETHODS $(CL, M', x_{Rj}, TC, t_{user}, c_{user})$
22:         **end for**
23:     **end for**
24: **end if**

For critical risk applications, our primary objective is to obtain the maximum CDC regardless of test cost and time. On the other hand, for low and very low risk applications, test time and cost are more important than getting the maximum CDC. For medium- and high-risk applications, we can get a higher confidence level by setting a higher test time and cost limit. For critical applications, the SORT() function (line 10) takes M and CL as arguments and sorts them according to x_{ij} and discards the method i when $x_{ij} = 0$. Equation (6.1) has been implemented by the CALCULATE() function (line 11). The SELECTMETHODS() function (line 13) takes x_{Rj} and TC as arguments and selects methods until the condition, $x_{Rj} \geq TC$, is met. If this condition is not met after iterating all the methods, then the defects belong to the $UCDs$. If $x_{Rj} = 0$, then the defects become $NCDs$.

For other applications, the SORT() function (line 18) takes M, T, and C as arguments and sorts according to linear combinations of t_i and c_i $(0.5t_i - 0.5c_i)$ and discards the method i when $x_{ij} = 0$. The resultant confidence level has been calculated by the CALCULATE() function (line 19) through the implementation of Eq. (6.1). The SELECTMETHODS() function (line 21) takes x_{Rj}, TC, t_{user}, and c_{user} as arguments and selects the methods that require the minimum test time and cost to achieve $x_{Rj} \geq TC$. If this condition is not met after iterating all the methods (as was the case for the critical applications, as well), then the defects belong to the $UCDs$ and, if $x_{Rj} = 0$, then the defects become $NCDs$.

6.4.2.2 Assessment of Test Methods

After the selection of the test methods, the assessment is done on those methods. It invokes Algorithm 1 with DI, DM, and selected methods by Algorithm 2 as inputs to compute CDC, CTC, $NCDs$, and $UCDs$.

6.4.2.3 Example

Let us now start with a simple example to explain the dynamic assessment. All the data used for this example are the same as the data used in Example 6.4.1.2.

Table 6.8 Sample example for the dynamic assessment

	Step	Name	Description
Selection (Algorithm 2)	Line 4–6	Read inputs	Read TL, CL, and DF
	Line 7	Sort DF	No sort needed as all the defects are treated equally
	Line 17–23	Select methods	Defect D1: Select method M1 Defect D2: Method M1 already selected and $x_{RD2} = TC$, No extra methods are necessary Defect D3: Select M2 Defect D4: No test methods can detect D4 Defect D5: Method M2 already selected and $x_{RD5} = TC$, No extra methods are necessary Selected methods are M1 and M2
Assessment (Algorithm 1)		Read inputs	Read DI, and DM
	Line 2–4	Compute resultant confidence (x_R) using Eq. (6.1)	$x_{RD1} = 1 - \{(1-0.9)(1-0)\} = 0.9$ $x_{RD2} = 1 - \{(1-0.5)(1-0)\} = 0.5$ $x_{RD3} = 1 - \{(1-0)(1-0.9)\} = 0.9$ $x_{RD4} = 1 - \{(1-0)(1-0)\} = 0.0$ $x_{RD5} = 1 - \{(1-0)(1-0.5)\} = 0.50$
	Line 5	Compute CDC using Eq. (6.4)	$CDC = 100 * \frac{1*0.9+1*0.5+1*0.9+1*0.0+1*0.5}{1+1+1+1+1+1}\% = 56\%$
	Line 6	Compute CTC using Eq. (6.7)	$CTC_x = 0.9 * \frac{1*0.9+0*0.5+1*0.9+1*0.0+1*0.5}{1+0+1+1+1} \times 100\% = 51.75\%$ $CTC_y = 0.5 * \frac{0*0.9+0*0.5+1*0.9+1*0.0+1*0.5}{0+0+1+1+1} \times 100\% = 23.3\%$ $CTC_z = 0.5 * \frac{1*0.9+1*0.5+0*0.9+0*0.0+1*0.5}{1+1+0+0+1} \times 100\% = 6.3\%$
	Line 7	Compute $NCDs$ using Eq. (6.8)	NCD: Defect D4 as $x_{Rd} = 0$
	Line 8	Compute $UCDs$ using Eq. (6.9)	None

Table 6.9 Assessment of test methods for low risk category (CDC, NCDs, and UCDs)

#	Test methods	CDC (%)	NDCs	UCDs
1	General EVI	35.4		
2	Internal inspection	49.7		
3	Bond pull	50.9		
4	Radiological inspection	64.2		
5	DC test at ambient temperature	75.7		
6	Key AC/switching parameters at ambient temperature	85.4		
	Test plan for low risk category	85.4	0	0

In this example, we will consider low risk categories. The target confidence (TC) corresponding to this risk category is 0.5 (described in Table 6.2). For simplicity's sake, we are not considering test time and cost in this example (Table 6.8).

6.4.2.4 Results

The dynamic assessment first recommends a set of test methods and then does an assessment of those methods. In this section, we are going to present (i) the best set of test methods from the complete set of current counterfeit detection test methods to provide the maximum CDC, and (ii) an optimum set of test methods from the complete set that maximizes CDC while also taking test cost, time budgets, and application test categories into account. As discussed in Sect. 6.4.1.3, inputs to the framework were obtained by consensus among subject matter experts and test labs involved in the G-19A group [12].

Table 6.9 shows the dynamic assessment of the test methods for the low risk category. Column 2 shows the recommended test methods. The test cost and time budgets are not mentioned here, as there is a confidentiality agreement between the CHASE Center and the G-19A group. Columns 3, 4, and 5 represent the CDC, $NCDs$, and $UCDs$, respectively. The first recommended test method, general EVI, contributes a coverage of 35.4 %, as before. The second recommended test method, internal inspection combined with general EVI, contributes 49.7 % coverage. The combined coverage of the total 6 test methods provides a final CDC of 85.4 %. We can see that there is a significant reduction in the total number of test methods (11 to 6) in the dynamic assessment. There are also no $NCDs$ or $UCDs$ for this risk category.

Table 6.10 shows the $CTCs$ for all the counterfeit types in the low risk category. Here we can observe the similar CTC values for the dynamic and static assessments for all the counterfeit types, as both assessments provide similar $CDCs$.

Table 6.10 Assessment of test methods for low risk category (CTC)

#	Counterfeit type	CTC
1	Recycled	84.4
2	Remarked	76.5
3	Overproduced	2.5
4	Out-of-spec./defective	80.1
5	Cloned	8.4
6	Forged documentation	62.9
7	Tampered	NA

Fig. 6.1 CDC for critical risk category

6.4.3 Comparison Between Static Assessment and Dynamic Assessment

The static assessment reports the test metrics when the user wants to perform the assessment on a fixed set of test methods. In the dynamic assessment, an optimum or a best set of test methods is recommended and then the assessment is performed. It is now necessary to compare the test results after performing these assessments. Here, we selected the test methods recommended by SMEs for all five risk categories for the static assessment, and provided all the test methods mentioned in Fig. 4.1 to the dynamic assessment for selecting an optimum set of test methods.

Figures 6.1, 6.2, 6.3, 6.4, 6.5 show the comparison of the increase in CDC for SMEs' and our proposed framework's recommended test methods for all five risk categories. The x-axis represents the number of test methods that result in the CDC, which is shown on the y-axis. The legend DA refers to the test methods recommended during the dynamic assessment. Figure 6.1 clearly shows that the rate of increase of CDC is not significant for a subgroup of test methods (e.g., test methods 2–7 do not add any CDC value to test method 1). This subgroup of test methods detects the same defects and does not contribute to the CDC. However, according to our framework, every test method contributed to the CDC, and we

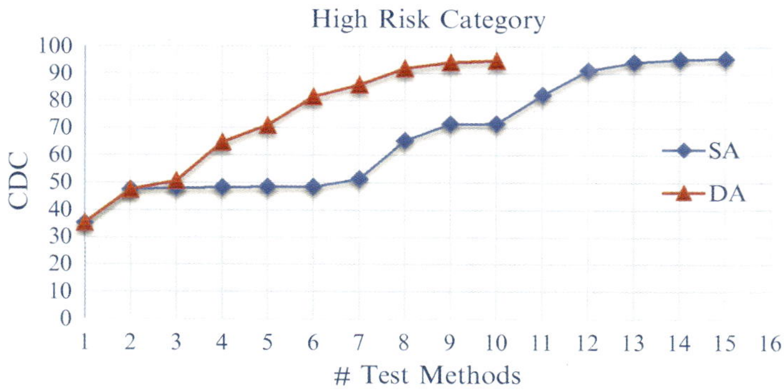

Fig. 6.2 CDC for high risk category

achieved 99.4 % of CDC by using only 14 test methods, whereas there are 19 test methods recommended by SMEs that provide the CDC of 99.1 %.

A similar trend is observed in Figs. 6.2, 6.3, 6.4, 6.5. We observed a significant reduction in test methods when using our framework to achieve a similar CDC with a much lower total test cost and time. The percentage reduction of test cost varies from 10 to 23 %, and test time varies from 10 to 24 % for different tier levels. We are not showing the actual values of test time and cost as there is a non-disclosure agreement between the CHASE Center and the G-19A group.

Table 6.11 shows the $CTCs$ for all the counterfeit types. Here we can observe the similar CTC values for the SA and DA recommended test methods for all six counterfeit types. However, DA provides improved $CTCs$ for overproduced, out-of-spec/defective, and cloned counterfeit types for low and very low risk categories, as the test methods recommended by DA provide much better coverage of defects related to electrical category.

Figure 6.6 shows the number of $NCDs$ and $UCDs$ for all five risk categories. Figure 6.6a illustrates that both sets of test methods result in the same $NCDs$ for critical, high, and moderate risk categories. For low and very low risk categories, the

Table 6.11 Assessment of test methods (CTC)

	Critical		High		Moderate		Low		Very low	
Counterfeit type	SA	DA	SA	DA	SA	DA	SA	DA	SA	DA
Recycled	97.1	97.4	93.5	92.8	92.5	91.6	82.8	84.4	73.2	76.6
Remarked	89.2	89.3	84.6	83.7	84.6	83.6	84.5	76.5	81.0	76.4
Overproduced	3.0	3.0	2.7	2.7	2.6	2.6	1.6	2.5	0.9	1.7
Out-of-spec/defective	96.7	97.3	89.4	88.9	85.7	85.0	53.3	80.1	29.8	54.6
Cloned	9.9	9.9	9.3	9.2	9.0	8.9	6.3	8.4	4.3	6.3
Forged documentation	68.9	68.9	68.9	68.8	68.9	68.8	68.9	62.9	68.8	62.9
Tampered	NA	NA	NA	NA	NA	NA	NA	NA	NA	NA

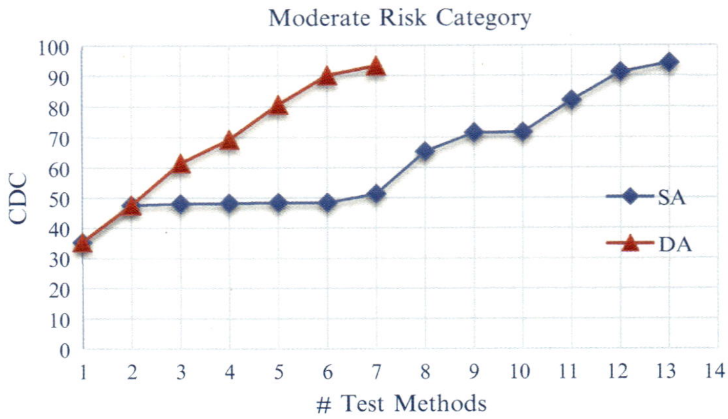

Fig. 6.3 CDC for moderate risk category

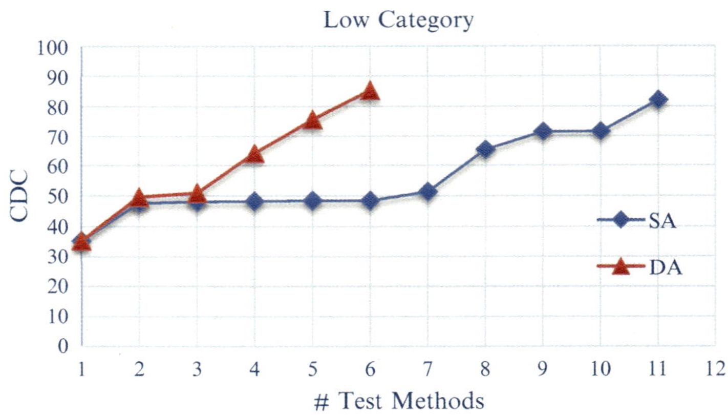

Fig. 6.4 CDC for low risk category

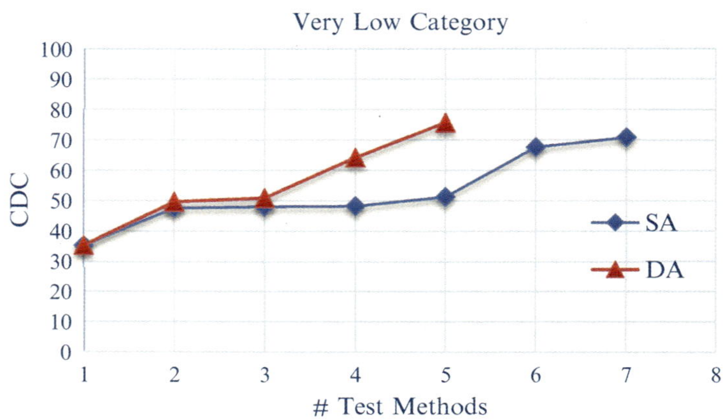

Fig. 6.5 CDC for very low risk category

DA recommended test methods provide much better coverage of counterfeit defects. The test methods recommended by SMEs for low and very low risk categories mostly emphasize the detection of procedural, mechanical, and environmental defects. Most of the electrical defects are undetected and we observe a high NCD value, especially for the very low risk category. In Fig. 6.6b, we observe similar UCD values for SA and DA-recommended test methods. For the low and very low risk categories, there are no $UCDs$, as the target confidence (see Table 6.2) is low and covered by the test methods.

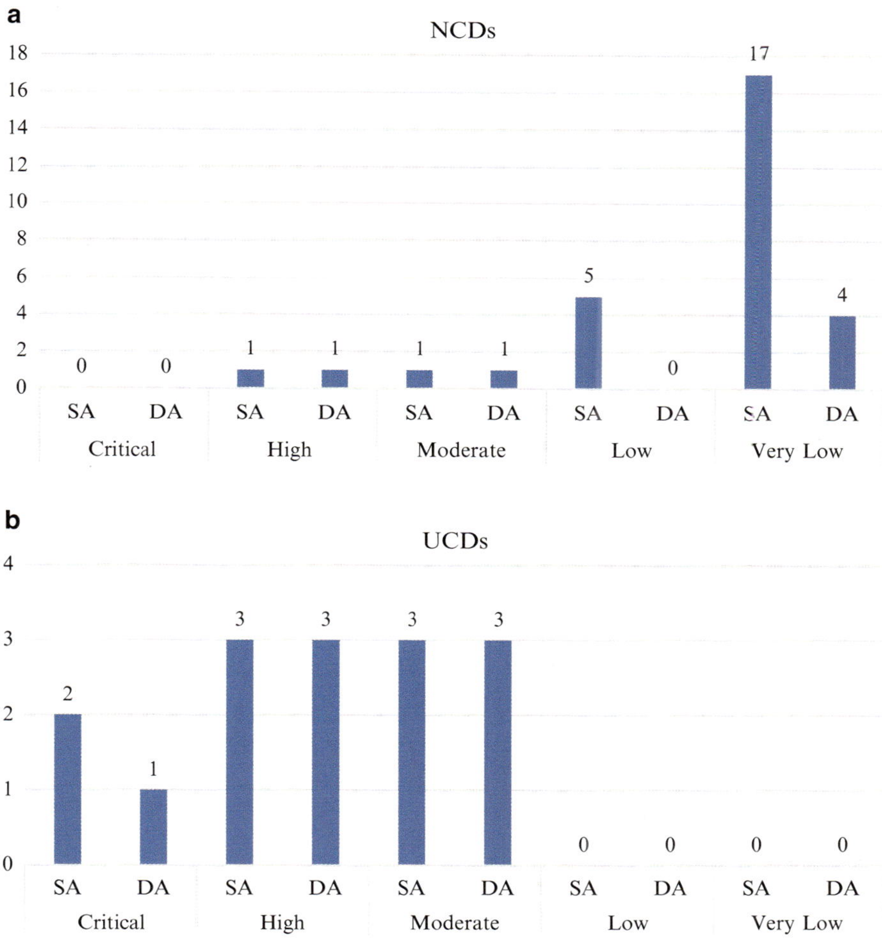

Fig. 6.6 Comparison study of NCDs and UCDs between the SMEs', and our framework recommended test methods (**a**) NCDs for all five risk categories, and (**b**) NCDs for all five risk categories

6.5 Summary

In this chapter, we have developed a comprehensive framework for assessing currently available test methods by introducing test metrics such as Counterfeit Defect Coverage (CDC), Counterfeit Type Coverage (CTC), Under-covered Defects (UCD) and Not Covered Defects (NCD). The framework provides two types of assessments: static assessment that helps in the evaluation of test methods based on the aforementioned metrics, and dynamic assessment for selecting an optimum set of test methods to minimize the test coverage. Static assessment can be used to estimate the counterfeit detection capabilities of a test lab based on their equipment and test methods. The dynamic assessment can be used by test labs to determine how much they can improve their capabilities by adding different equipment and test capabilities. It can also illustrate the trade-off between test time, cost, and counterfeit coverage. Both types of assessment can determine what counterfeit defects are partially covered or missed, what counterfeit types are not well covered, etc. This information can be used to guide in the development of new test methods for counterfeit detection.

The challenge of collecting data still remains a problem. In this chapter, we pointed out that the data used as inputs to the framework (such as confidence level matrix, decision index etc.) implemented by UConn's CHASE center is based on compiled information from subject matter experts and participating labs in the G-19A group. There is a need for such data to be collected at a larger (or even global) scale to enhance test coverage. The first step of collecting data will be addressed in Chap. 7. Currently, adequate coverage exists for recycled, remarked, out-of-spec/defective, and forged documentation counterfeit types. The low coverage for overproduced and cloned counterfeit types indicates that there is an urgent need for developing new design-for-anti-counterfeit (DFAC) measures to address the effective detection of these counterfeit types.

References

1. U. Guin, D. DiMase, M. Tehranipoor, Counterfeit integrated circuits: detection, avoidance, and the challenges ahead. J. Electron. Test. **30**(1), 9–23 (2014)
2. U. Guin, K. Huang, D. DiMase, J. Carulli, M. Tehranipoor, Y. Makris, Counterfeit integrated circuits: a rising threat in the global semiconductor supply chain. Proc. IEEE **102**(8), 1207–1228 (2014)
3. U. Guin, M. Tehranipoor, D. DiMase, M. Megrdichian, Counterfeit IC detection and challenges ahead. ACM/SIGDA E-NEWSLETTER **43**(3) (2013)
4. U. Guin, D. Forte, M. Tehranipoor, Anti-counterfeit techniques: from design to resign, in *Microprocessor Test and Verification (MTV)* (2013)
5. U. Guin, M. Tehranipoor, On selection of counterfeit IC detection methods, in *IEEE North Atlantic Test Workshop (NATW)* (2013)
6. U. Guin, D. DiMase, M. Tehranipoor, A comprehensive framework for counterfeit defect coverage analysis and detection assessment. J. Electron. Test. **30**(1), 25–40 (2014)

7. CHASE, ARO/CHASE Special Workshop on Counterfeit Electronics, January 2013, http://www.chase.uconn.edu/arochase-special-workshop-on-counterfeit-electronics.php
8. CHASE, CHASE Workshop on Secure/Trustworthy Systems and Supply Chain Assurance, April 2014, https://www.chase.uconn.edu/chase-workshop-2014.php
9. SAE, Test Methods Standard; Counterfeit Electronic Parts. Work in Progress, http://standards.sae.org/wip/as6171/
10. ERAI, Report to ERAI, http://www.erai.com/information_sharing_high_risk_parts
11. GIDEP, How To Submit Data, http://www.gidep.org/data/submit.htm
12. G-19A Test Laboratory Standards Development Committee, http://www.sae.org/servlets/works/committeeHome.do?comtID=TEAG19A

Chapter 7
Advanced Detection: Physical Tests

In Chap. 4, we identified the challenges and limitations of common physical inspection methods. It was noted that common counterfeit detection practices can fail to detect more and more sophisticated counterfeiting. Counterfeiters are utilizing more advanced techniques making the discrepancies from the authentic ICs so subtle and at times impossible to detect. This in turn calls for more advanced detection techniques to keep up with the counterfeiters' pace. We also discussed that current detection practices rely on subject matter experts (SMEs) to interpret the results of the characterization techniques which eliminates the possibility of effective automation in addition to creating inconsistencies. These challenges along with the destructive nature of several common detection techniques urge researchers to think of novel ideas which can provide more effective detection practices.

This chapter will focus on two novel characterization methods which can provide multidimensional information of integrated circuits: Four-dimensional Scanning Electron Microscopy and three dimensional (3D) X-ray Microscopy. It will be shown that simple two dimensional (2D) imaging currently utilized in detection lacks the level of sophistication required to detect more subtle defects. The novel methods will then be introduced and their effectiveness in detecting previously missed defects will be demonstrated. Finally, to address the issue of automation, we first address the issue of quantification. As mentioned earlier in Chap. 4, the lack of consistent metrics in counterfeit detection have left the community with no choice but to rely on subject matter experts. Here, several new statistical parameters are introduced to address one of the most challenging defects: Improper texture or texture variations. Such defects are usually consequences of sanding, remarking or resurfacing which are believed to be some of the most frequent but yet challenging to detect phenomena in counterfeit ICs.

M. Tehranipoor et al., *Counterfeit Integrated Circuits*,
DOI 10.1007/978-3-319-11824-6_7

7.1 Limitation in 2D Characterization

Traditional optical, digital and electron microscopy provide us with two-dimensional (2D) information. While these methods can effectively detect simple, obvious defects (e.g., dents, incorrect dimensions, etc.), they are insufficient for more complex defects such as improper texture and dimple height variation, especially if the counterfeiters perform resurfacing at very high quality. Figure 7.1 shows instances where traditional 2D methods 7.1a can and 7.1b cannot detect resurfaced components. Figure 7.1c, d show how the 3D data can detect resurfacing in the component through presence of extra material at the wall of the dimple and also texture variation.

Blacktop coating, a frequent phenomenon in counterfeit ICs, is barely distinguishable using conventional Scanning Electron Microscopy (SEM) and it depends on the acquisition parameters on the instrument. Figure 7.2 shows two SEM images of the exact same location obtained by different acquisition parameters. While both images have optimized brightness and contrast values, the variation in the texture is almost entirely invisible in the image on the right; the left image on the other hand

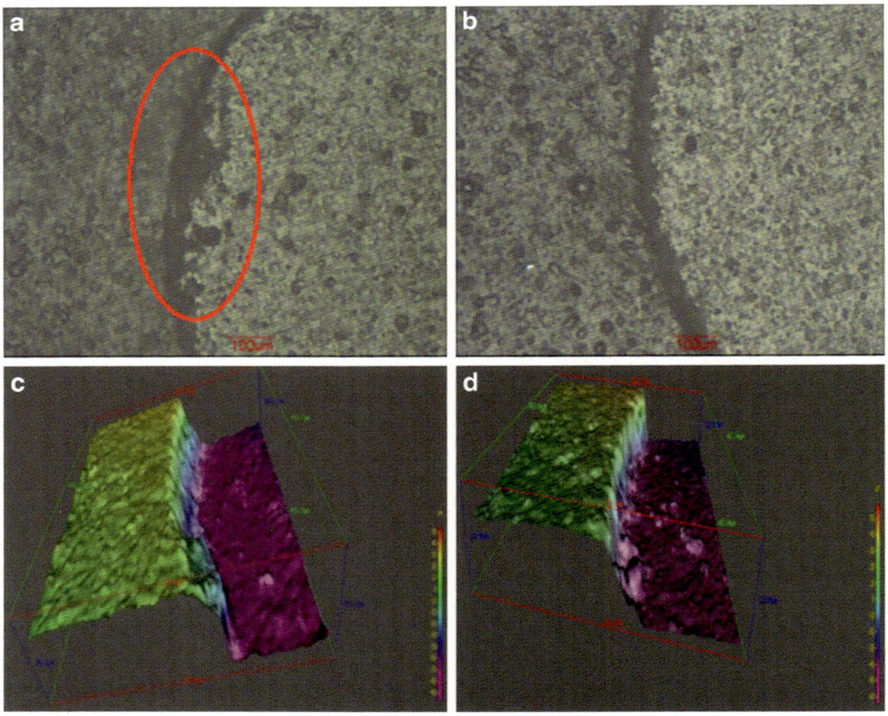

Fig. 7.1 Evaluation of a dimple on microchip package at 2D (**a**, **b**) and in 3D (**c**, **d**)

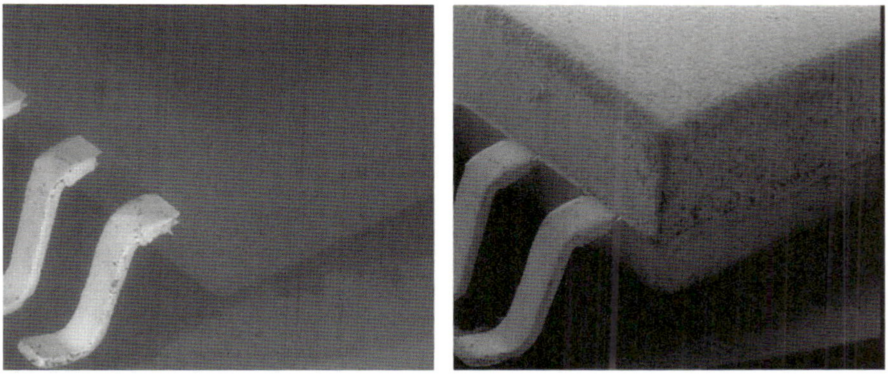

Fig. 7.2 SEM images of the exact same location of a known top coated IC with blacktop coating invisible in one (*left*) and detectable in another (*right*)

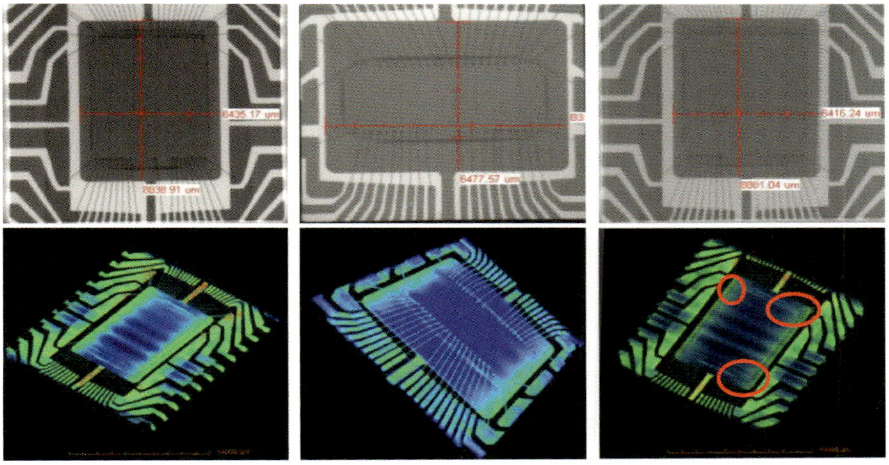

Fig. 7.3 2D vs. 3D investigation—internal die-related defects

clearly shows presence of a thin coating layer applied on top. Furthermore, such images always require an SME to interpret the data as it lacks quantified information and consequently is not suitable for automation.

The limitation of two dimensional information is not limited to the exterior of the ICs and the texture-related defects. One can find similar limitations in 2D images obtained from the inside of the microchips using conventional radiography.

The present practices in industry propose 2D X-ray imaging as a mandatory and preliminary test [1] to uncover interior defects associated with die and bond wires. However, in our experience we have found that several defects remain undetected without the third dimension. In Fig. 7.3, we demonstrate three samples that share same exterior appearance, lot code, and markings; however, one is authentic and two are counterfeit (shown respectively in first, second and third columns). The 2D

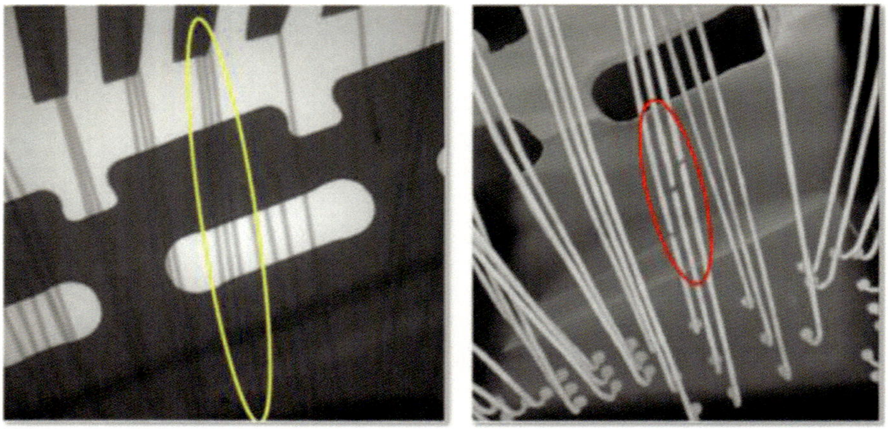

Fig. 7.4 2D (*left*) vs. 3D (*right*) bond wire imaging [5] (http://www.zeiss.com/corporate/en_us/home.html)

X-ray radiography (first row on Fig. 7.3) can detect that the sample in the second column has different die orientation and dimensions suggesting that it has been remarked, but the same test for the sample in the third column is inconclusive.

In the second row of Fig. 7.3, we have demonstrated how 3D images reconstructed from X-ray tomography can reveal quality concerns in the third sample. Specifically, we have highlighted areas along the die face that are delamination of the molding compound. Die face delamination is a threat to reliability because they can expand during use until they shear off bond wires. The only other alternative that the community resorts to for detecting such defects, is using scanning acoustic microscopy (SAM) [2, 3]. Though SAM is typically considered non-destructive, it requires that the sample be immersed in water. Water exposure can harm or possibly destroy the chip. Other possible ways, involve the decapsulation of the chip which are inherently destructive [4]. We demonstrate another example of insufficiency of 2D information where the defect involves the bond wires.

In Fig. 7.4, the 2D information is not sufficient for detection of fine cracks. Such information can be critical in estimating the reliability of chips. Other instances of the insufficiency of 2D radiography can be found in [5, 6].

Finally, as earlier discussed in Chap. 4, material composition characterization is a valuable tool that can not only improve the possibility and confidence of counterfeit detection but also enhance the capability of detecting counterfeit ICs to more counterfeit types where means to detect such counterfeit types are very limited. For instance, comparing the material composition of the component's top, back and side can more often yield conclusive evidence of the samples being resurfaced or blacktop coated.

There are a variety of techniques possible for attaining the above characterization ranging from X-ray fluorescence (XRF) [7] to Raman spectroscopy and FTIR [8] but many of them are restricted by the following: (i) they are mainly limited to the surface of a sample and cannot go beyond few nanometers in depth; (ii) the elemental analysis is mostly done in one dimensional fashion providing average result of the entire scanned area, while a thorough examination requires location of an element in addition to its presence; (iii) compositional analyses are collected separately and presented in the form of a spectrum rather than an image which can hinder the automation of the counterfeit process; (iv) The analysis usually involves multiple imaging sessions increasing cost and technician time.

The following sections of this chapter introduces how the novel and more advanced characterization methods can address the above mentioned issues of conventional detection methods.

7.2 Four Dimensional Scanning Electron Microscopy

Scanning electron microscopes are extensively used in material science as a characterization tool. SEM micrographs have been qualitatively used to characterize the surface geometry of a specimen using the secondary electron (SE) mode; however, these images are essentially two dimensional, lacking quantitative depth information. Since the early days of the availability of SEMs, efforts have been made to extract the three dimensional (3D) information from SEM images [9–16]. SEM-based 3D imaging offers major advantages compared to other methods. SEMs have a large depth of field [13, 16], which makes it possible to have features at radically different heights in focus simultaneously. The lateral resolution and signal to noise ratio in SEM images are also much better than optical methods [13]. In addition, there is no mechanical contact in SEM imaging, which makes it a proper instrument for very rough surfaces [16] with steep valleys and overhangs—such as some mold marks (dimples) on integrated circuits. Several SEM-based 3D imaging techniques have already been developed, such as Focus Ion Beam (FIB) tomography [17], shape from shadows [18, 19], and stereo-photogrammetry [9–16], the latter of which will be used in this novel method. The stereo-photogrammetry technique requires computationally intensive image processing, but the presence of fast computers in recent years has resulted in more interest in this method [20]. A general stereo-photogrammetric technique consists of three stages [16, 20–22]:

i. *Acquisition stage*: Acquires images of the same area of the specimen at 2 or more perspectives
ii. *Matching stage*: Matches the acquired images to find points corresponding to the same position
iii. *Depth extraction stage*: Extracts the third dimension based on projection geometry to obtain the 3D model

There are certain challenges and difficulties at each of these stages which can bring uncertainty to the final reconstruction result [13, 20, 23]. Here, each step is explained in detail. In addition, certain procedures are introduced to reduce the errors at each step and produce more reliable and repeatable quantitative results. Also, back-scattered secondary electron (BSE) mode of an Everhart–Thornley detector [24] (ETD) has been used for the first time for the high fidelity reconstruction, especially where SE images suffer from charging. Finally, a modified procedure is introduced to quantitatively record the surface geometry of any SEM sample. SEM photogrammetry is already used in various applications ranging from material science [16, 25, 26] to dentistry [14, 15], and the availability of these new procedures can enhance the fidelity of such measurements as well.

The challenges associated with each of the three stages discussed above and suggested remedies to overcome them are fully explained in the subsections below. The effect of the remedies are also explained and quantitatively demonstrated.

7.2.1 Acquisition Stage

At this stage, SEM-micrographs are to be obtained from different perspectives of the same area. Since the detectors are stationary in an SEM instrument, multiple perspectives are obtained either by eucentrically tilting the stage or by rotating the sample which has already been mounted with a tilt angle on the SEM stub and choosing a correct scan rotation. The latter technique is applicable in cases where the instrument has some restrictions for maximum negative and positive tilt. Note that the images shown in this chapter are acquired using an FEI quanta FEG 450 SEM machine which does not have such limitations. We have also had experience in using the JEOL 6335FESM where the maximum negative tilt angle could not exceed 5°. The tilt angles, in both positive and negative direction, have to be at least +5 and −5° [27], and in the case of flat objects, larger tilt angles can yield better results [20]. Therefore, in cases where the required tilt angle is larger than the instrument limit, rotation of an already tilted sample is more effective. Other ways of acquiring images at different perspectives are also available [23], but tilting has been the easiest one [20], and therefore has been adopted in this chapter.

For extraction of height information, one needs at least two images at different viewing angles. However, in order to perform an image processing-based automatic calibration of tilt angle, a third image at 0° is suggested [27]. There are also software packages available for performing digital photogrammetry [28–30].

During the image acquisition stage, certain requirements have to be met:

1. Images must have appropriate illumination with no charging and no stripes.
2. The tilting has to be eucentric: that is, the center of images at different perspectives has to be the same.
3. The magnification and working distance have to remain the same for all images and must be recorded for reconstruction.

Maintaining these requirements at the same time results in certain challenges which are addressed below.

7.2.1.1 Image Quality Challenge

It is absolutely necessary that images be obtained at proper illumination, avoiding any excessively bright and dark spots [20, 27], which are almost exclusively due to charging. Such defects, if minimal, are usually rectified by adjusting the contrast and brightness, shortening the scan dwell time, or using low vacuum or Environmental SEM (ESEM) modes, where the charging is attributed to the presence of nonconductive materials on the sample. In cases of excessive charging, even in a low vacuum and an ESEM, a conductive coating of gold or carbon is usually applied on the sample to avoid charging [31]. Though fairly effective, such coatings may be hard to remove from ICs with deep valleys and holes, which could be destructive to the sample. Most images in this chapter are produced using the BSE mode of an ETD detector where charging could be avoided. In presence of excessive charging the low vacuum mode with variable pressure was utilized to remove the charge.

7.2.1.2 Challenges During Tilting

The challenges related to tilting, their causes, and their remedies are summarized in Table 7.1 and are explained below:

A) Tilting axis

Figure 7.5a shows an image of the inside of the FEI quanta FEG 450 instrument taken by a built in CCD camera. The coordinate system drawn on the image illustrates the possible motion of the stage. This particular SEM can move along any of X, Y, Z axes. In addition, the stage allows rotation about Z and tilting about Y.

Table 7.1 Tilting challenges, causes and remedies

Challenge	Reason	Proposed remedies
Axis of tilt being different from the one being used in data reduction software	Instrument restrictions; Possibility of hitting a detector during tilting	1. Rotating the images by 90° (previously used)[27] 2. Using scanning rotation of 90° (much more effective)
Shift of the center of image during tilting	Non-ideal tilting stage	Reorienting the center of the image after tilting
Adjusting loss of focus during magnification,without a change of working distance (must be at high magnification to notice)	Change of relative height of the center of the sample	Refocusing by stage height adjustment in tilted Z axis (at high magnification)
Uncertainty in exact tilting angle	Precision restriction on the instrument	Image Processing on CCD images of the inside of the instrument

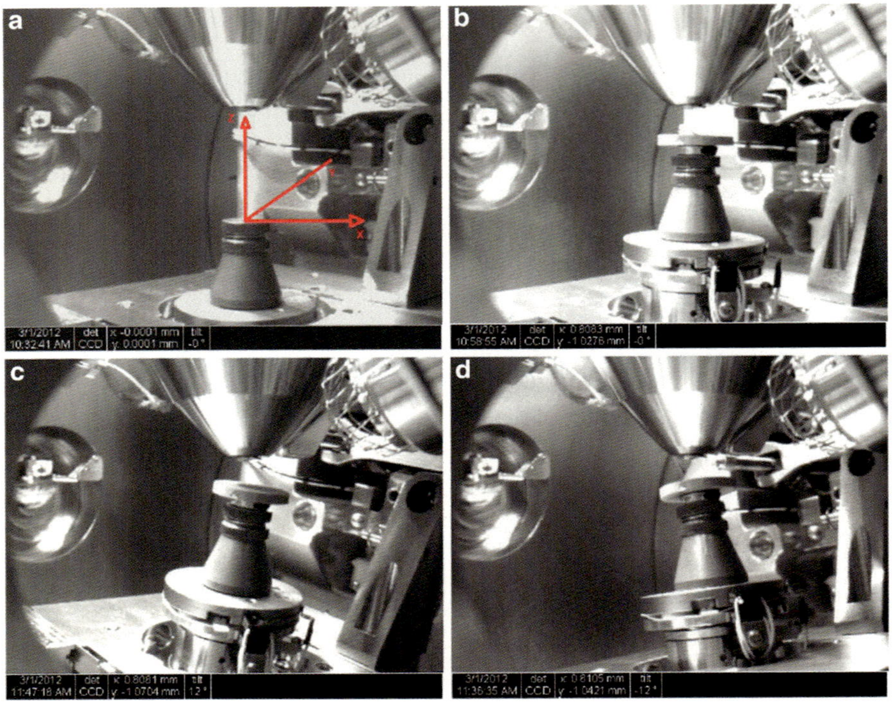

Fig. 7.5 CCD images of the inside of an SEM instrument (**a**) SEM stage with the assigned coordinate system, (**b**) 0° tilted sample inside SEM chamber, (**c**) +12° tilted sample inside SEM chamber, and (**d**) −12° tilted sample inside SEM chamber

This many of degrees of freedom can be found in almost any SEM, with the only difference being the axis of tilt. In some SEMs the X axis and others the Y axis is chosen for tilting the stage. The algorithm used for extracting height introduced later in this chapter and also preferred by many available software packages, is based on the axis of tilt being parallel to the vertical axis of the image plane [12]. A simple procedure to recognize this is to see whether features are shifted horizontally or vertically during tilting. If features shift vertically, that is, the axis of tilt is parallel to the horizontal axis of the image plane; the image is not compatible with the algorithm. A previously proposed remedy for such an issue was rotating the images by 90° before reconstruction [27]. Although such a remedy can yield results, it has a negative effect on the resolution and field of view of the reconstructed surface. A clockwise 90-degree scan rotation is suggested here as an alternative remedy. Using this modification, the images are rotated during the scan rather than afterwards, which results in better repeatability and bigger field of view. Table 7.2 shows this improvement at 500X magnification from our reconstruction experiment described subsequently.

Table 7.2 Effect of remedies for horizontal axis of tilt

Height variation using image rotation	Height variation using scan rotation	Lateral length field of view using image rotation	Lateral length field of view using
365 nm	340 nm	270.02 μm	300.42 μm

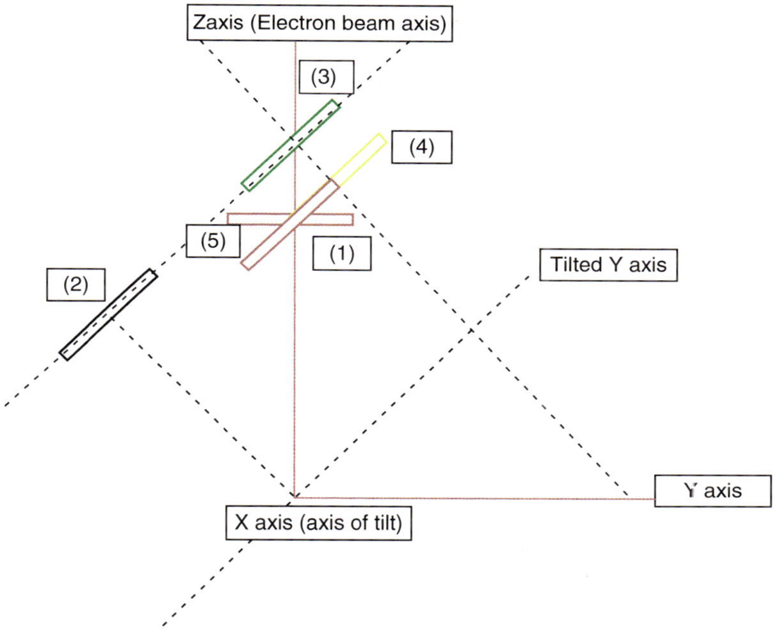

Fig. 7.6 Schematic of the sample movement during the tilt. The figure is exaggerated for better visualization

Figure 7.5b–d demonstrate the orientation of the sample at untilted, positive tilting angles, and negative tilting angles respectively and at the same working distance.

B) Center shift and focus correction

During tilting, the sample will be shifted horizontally or vertically depending on the tilt axis. Figure 7.6 shows a schematic of the sample movement during tilting.

The sample is initially at position (1) when the stage has not been tilted. After the tilt, the stage will be at position (2) (horizontal shift), which requires us to shift it back to the center. Position (3) shows the sample after the shift. However, now the sample is relatively out of focus, which can easily disturb 3D image reconstruction. Typically, the focusing would be readily achieved by changing the working distance; however, since the change of working distance changes the magnification and distorts calibration, the sample has been carefully moved along the tilted Z-axis to be back in focus (Position (4)). The focusing process results in some vertical and horizontal shifting which has been rectified by movement along the corresponding axes. Therefore at the final position (5), the center of the image will be back to its previous location before the tilt.

7.2.2 Depth Extraction Stage

The depth information is found using the modified Piazzesi algorithm [12]. Figure 7.7 demonstrates the projection geometry. Mex software [27] is used to do the computation automatically.

Any point $P(x, y, z)$ on the sample has the projection $P'(x', y')$ in the 2 dimensional plane of XY. Using Polar coordinates in YZ plane, we can write:

$$y = R\, cos\theta$$

$$z = R\, sin\theta$$

Based on trigonometry relations and Fig. 7.7, we can associate the x' and y' coordinates to the projection distance d, R, and θ:

$$x' = \frac{x}{1 - (R/d)\, sin\theta}$$

$$y' = \frac{R\, cos\theta}{1 - (R/d)\, sin\theta}$$

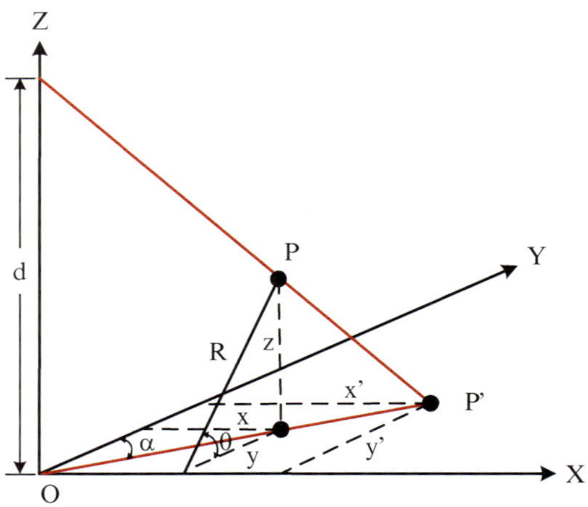

Fig. 7.7 Geometry of projection using Piazzesi algorithm

Following any tilt by $\Delta\theta$ we can write the new x' and y' as below, where the subscripts show the value at each step.

Step 1: (Negative tilt)

$$P(\theta - \Delta\theta) : x_1 = \frac{x}{1 - (R/d_1)\, sin(\theta - \Delta\theta)}$$

$$y_1 = \frac{R\, cos(\theta - \Delta\theta)}{1 - (R/d)\, sin(\theta - \Delta\theta)}$$

Step 2: (Positive tilt)

$$P(\theta + \Delta\theta) : x_1 = \frac{x}{1 - (R/d_1)\, sin(\theta + \Delta\theta)}$$

$$y_1 = \frac{R\, cos(\theta + \Delta\theta)}{1 - (R/d)\, sin(\theta + \Delta\theta)}$$

And finally having two tilted points by solving the algebraic equations, the coordinates of any point P can be calculated as below:

$$z = \frac{(y_1 - y_2)cos(\Delta\theta) + y_1 y_2(1/d_1 + 1/d_2)sin(\Delta\theta)}{sin(2\Delta\theta)(1 + y_1 y_2/d_1 d_2) + cos(2\Delta\theta)(y_1/d_1 - y_2/d_2)}$$

$$x = \frac{d_1 + d_2 - 2z cos(\Delta\theta)}{d_1/x_1 + d_2/x_2}$$

$$y = \frac{z\big((y_1 + y_2)cos(\Delta\theta) + (d_1 - d_2)sin(\Delta\theta)\big) - (y_1 d_1 + y_2 d_2)}{(y_1 - y_2)sin(\Delta\theta) - (d_1 + d_2)cos(\Delta\theta)}$$

Following this stage, the 3D information of the surface is available and can be visualized as a 3D surface using triangulation. Figure 7.8a and b demonstrate the acquired images on a dimple of an IC taken at two tilting angles of positive and negative ten degrees. The working distance and the angle has been optimized using the above guidelines. Figure 7.8c shows the 3D reconstructed image with the color map corresponding to height values.

In order to extend the information to the 4th dimension, EDS (energy dispersive spectroscopy) mapping has been used on the exact same area and the images are reconstructed in color to show any variations in material composition. Figure 7.9 demonstrates a similar image to that of Fig. 7.8a and b in terms of location, however in this figure color represents material composition and shows the presence of different material on the wall of the dimple.

Presence of inorganic materials on the package of these chips is an anomaly. Further analysis has shown the same material composition, that is (Ti and V), on the markings of the chip. The presence of Si in the image is also an indication of sanding. This analysis can help us to not only distinguish the defect but also find out the process of counterfeiting which can contribute to further counterfeit detection

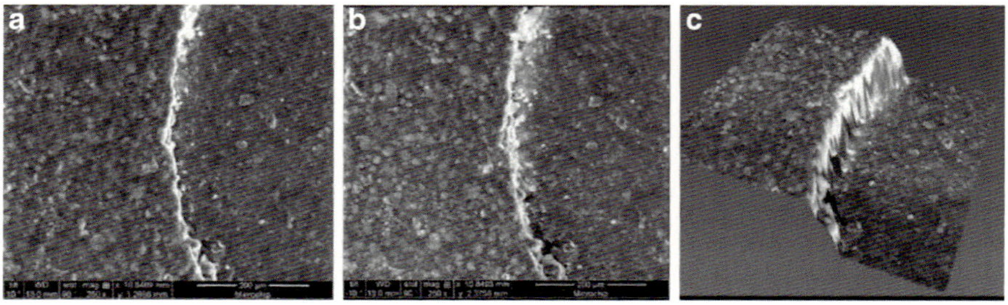

Fig. 7.8 Tilted (**a**) and (**b**) and reconstructed 3D (**c**) images

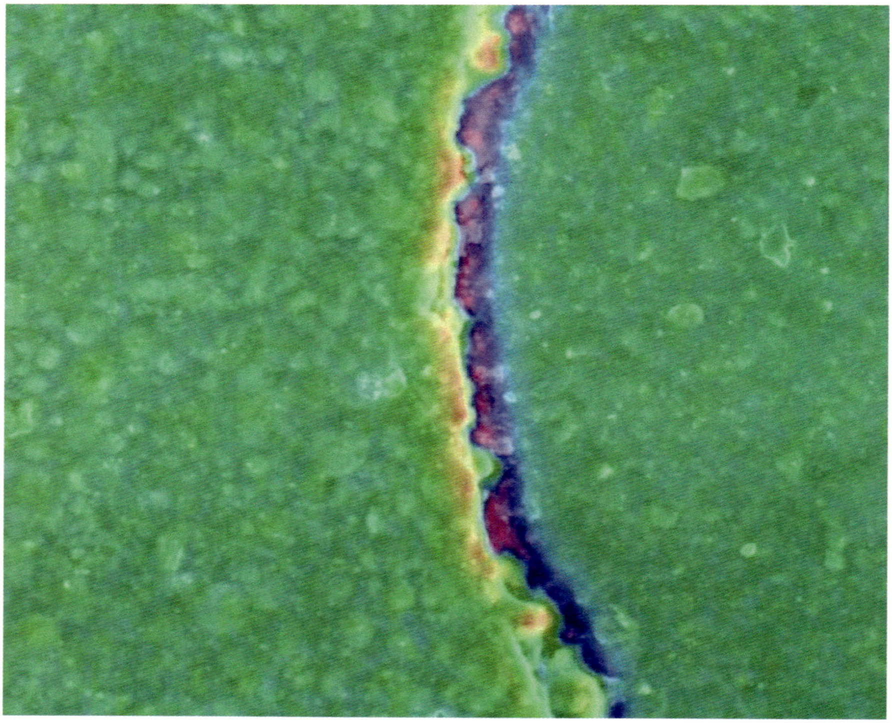

Fig. 7.9 Compositional map on the dimple, where *green* represents C and *blue*, *red* and *yellow* are Si, Ti and V respectively

and prevention. In this case, it appears that the previous marking on the chip has been removed using SiC paper sanding and then remarked. Combining Figs. 7.8c and 7.9 can provide us with four dimensional information on the sample which can greatly facilitate the counterfeit detection and is shown in Fig. 7.10.

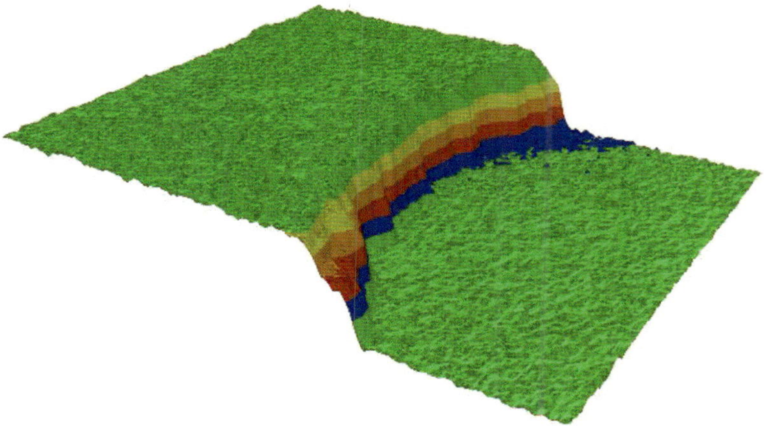

Fig. 7.10 Four-dimensional image of a dimple, with *colors* represent sample material as Fig. 7.9

Table 7.3 Surface parameters for texture analysis

Parameter	Definition	Formulation
Sa: Average roughness	Arithmetic mean of absolute height values of the area	$\sqrt{\frac{1}{MN}\sum_{\substack{1\le i\le M\\ 1\le j\le N}} Z(i,j)}$
Sq: RMS roughness	Root mean square value of the height of the area	$\sqrt{\frac{1}{MN}\sum_{\substack{1\le i\le M\\ 1\le j\le N}} Z^2(i,j)}$
Sp: Peak	Largest height of the area	$\max_{\substack{1\le i\le M\\ 1\le j\le N}} Z(i,j)$
Sv: Valley	Minimum height of the area	$\min_{\substack{1\le i\le M\\ 1\le j\le N}} Z(i,j)$
Sku: Peakedness	Kurtosis of the area	$\frac{1}{MNSq^4}\sum_{\substack{1\le i\le M\\ 1\le j\le N}} Z^4(i,j)$

7.3 Quantification of a 3D Surface: Improper Texture Variations

The data density and three dimensional information associated with 3D SEM allows to perform a more detailed texture analysis. Table 7.3 shows the surface parameters that are extracted from the surface of the chips to quantitatively address inconsistencies. In the table, Z denotes the height information that can be extracted through the entire rectangular surface with the length M and width N.

Roughness parameters such as Average and RMS allow us to have a better understanding of texture variations. Height parameters can help us analyze the

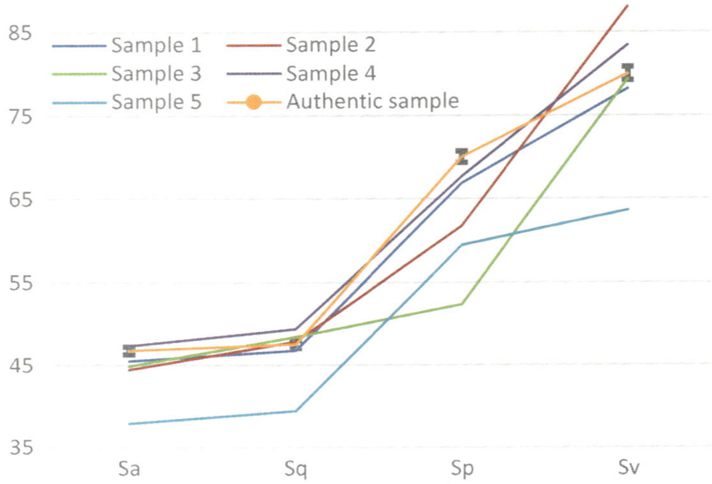

Fig. 7.11 Surface parameters extracted from a dimple for the five samples in study (all values are in microns)

dimple height variations independent of the reference plane and finally kurtosis, as a measure of pointiness of the surface allows us to find sanding marks and removed markings based on the texture information. Figure 7.11 shows the surface parameters extracted from the 5 ICs showing that using any single parameter alone cannot show the discrepancy among the surfaces however as more surface parameters are used. The inconsistencies among the samples are better identified.

It can be seen at roughness level (Sa and Sq), only sample 5 is showing discrepancy. Moving to height information we can see that sample 1 and sample 3 share different values than those of 2 and 4. This take us to our sixth measure, Kurtosis which has been kept out of the Fig. 7.11 for scaling issues. The kurtosis values for the 5 samples are 0.98, 1.5131, 1.3048, 1.5461, and 1.358.

This suggests that the value is substantially lower for sample 1 which can be attributed to sanding. The 4-dimensional image shown in Fig. 7.10 also showed sanding residual materials which can further prove the process of counterfeiting.

Such characterization allows us to quantitatively prove sanding which can facilitate the process of automation. In addition, as only inconsistencies within the five samples are used for detection, the analysis does not require a "golden IC". However, we have included the information about the authentic samples for validation purposes. As one can determine:

i. Whether such variations also appear different authentic samples
ii. Whether the samples have different surface parameter values than the golden IC.

The orange line in Fig. 7.11 and its error bars show the parameters for authentic samples and their variations among five samples. It is clear that the variations between the authentic samples are much smaller than that seen among the five studied counterfeit ICs. Also we can see roughness parameters are less effective measure comparing to height parameters and Kurtosis values.

Additional texture analysis using 3D SEM data is calculating the areal autocorrelation function (AACF) of the surface using the Eq. (7.1) to obtain the texture directionality [12, 32, 33]:

$$R(t_i, t_j) = \frac{1}{(M-i)(N-j)} \sum_{l=1}^{N-j} \sum_{k=l}^{M-i} Z(x_k, y_l) Z(x_{k+i} y_{l+j}) \tag{7.1}$$

where, $i = 0, 1, \ldots m < M; j = 0, 1, \ldots n < N; t_i = i.\Delta x; t_j = j.\Delta y$

Figure 7.12 shows the AACF of each of the surfaces where the value of autocorrelation ranges from 0 (pink—no correlation) to 1(red—maximum correlation) shown in color maps. Additional information on AACF can be found in references [34–37]. The ideal uniform direction in sample 1 also proves the sanding case. In addition further discrepancies were recorded between samples that initially had relatively similar roughness parameters.

To further investigate if similar cases can be identified in authentic samples, similar studies were conducted on five authentic samples. All figures and exact values have been excluded for brevity. One has been shown in Fig. 7.12 as an example. All the authentic samples shared a similar ACCF. This further proves that discrepancies in the AACF can be used as another metric for counterfeit detection.

7.4 3D X-Ray Microscopy

2D radiological X-rays is a well-established technique to non-destructively examine the interior attributes of the chip [38]. By taking multiple 2D X-ray projections and then reconstructing them, one can obtain the three dimensional information of the interior and the exterior image of the imaged object. This technique is commonly known as X-ray Tomography scan. Researches have used the 3D X-ray Tomography for studying electronic parts, however there have always been some issues [33]:

1. Many electronic parts have high aspect ratio, which results in the obligation to have a sufficiently large working distance during tomography to avoid collision of the sample with the source and detector. However, conventional CT loses resolution significantly as the distance to the source is increased. The number of X-ray counts also decrease greatly resulting in a much worse signal to noise ratio.
2. In many cases such as the ICs studied in this chapter, samples are made of materials with radically different X-ray absorption coefficients. For example, the package is commonly made out of Carbon with a very low X-ray attenuation

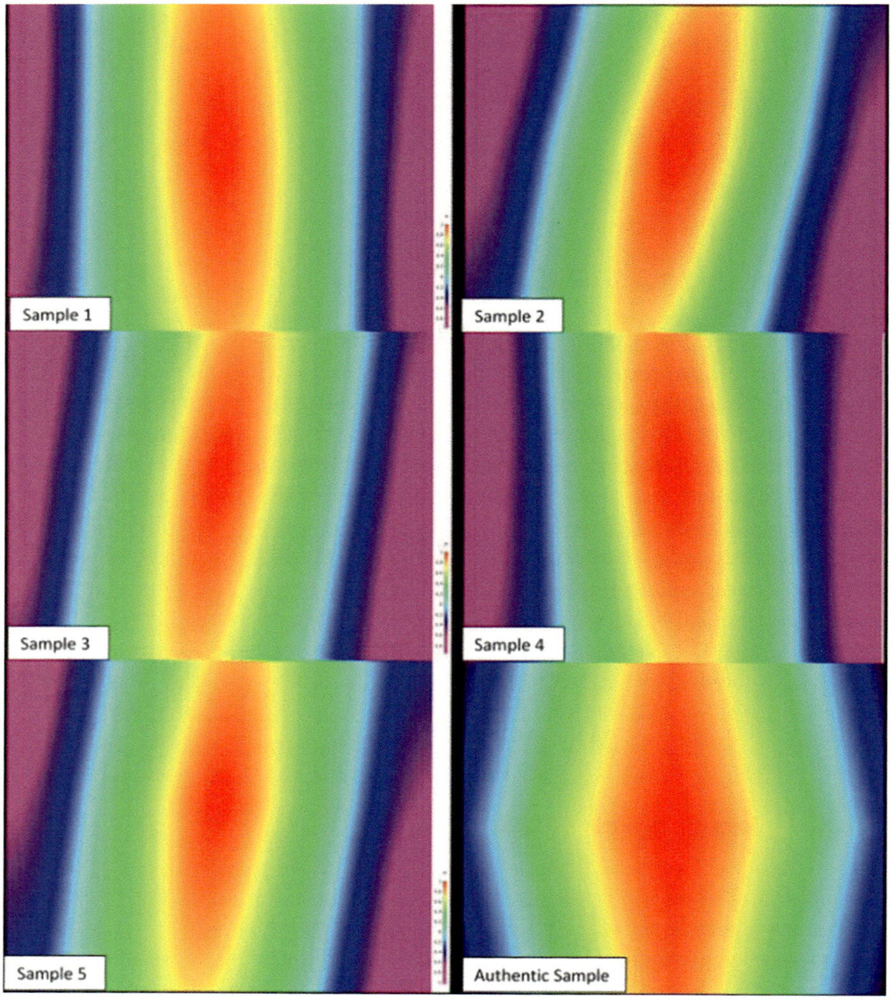

Fig. 7.12 AACF of all samples showing different texture direction

coefficient while the leads are made of Tin (Sn) which is a metal with a much higher attenuation coefficient. This entails problems associated with choosing the right X-ray source energy.

3. The process of tomography is believed to be time consuming and inefficient for everyday counterfeit detection application.

In order to overcome these challenges, two tomographies have been performed at two different energy levels focusing on the package for the low energy ones and investigating leads with a higher energy. Use of dual energy can help us to identify

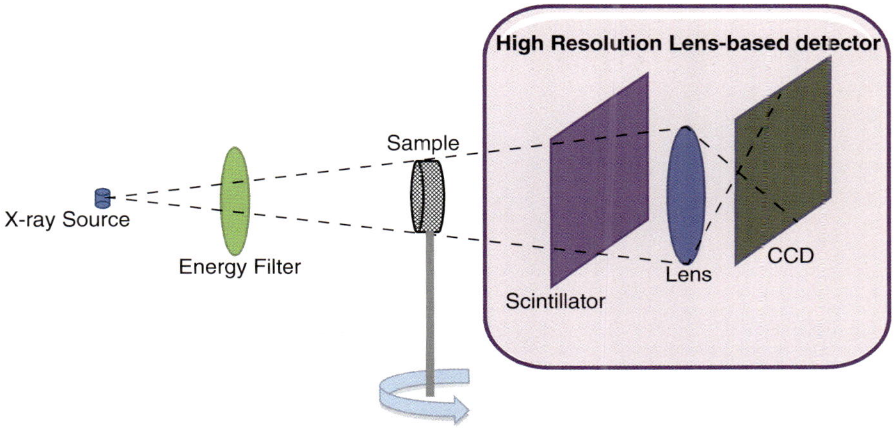

Fig. 7.13 Schematic of the X-ray microscope structure

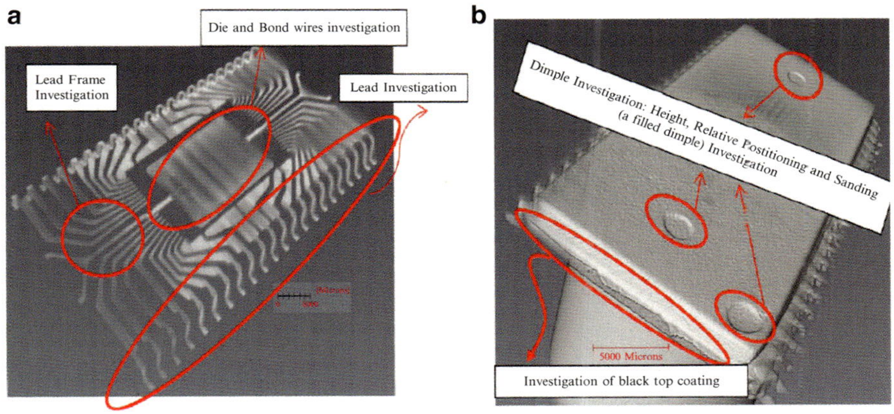

Fig. 7.14 XCT images of ICs using low (**a**) and high (**b**) energy

both exterior and interior defects. Also, as it is shown in the microscope architecture in Fig. 7.13, the use of a scintillator allows a much better resolution to be achieved at higher working distances. Figure 7.13 provides a schematic of the X-ray microscopy with resolution at a distance (Raad) propriety capability.

Also to capture the most defects in a single imaging session, the large field of view detector has been used where the entire sample can be imaged at once. Figure 7.14 shows a sample 3D rendered image at both high (a) and low (b) energies and the investigations that have been facilitated using the proposed technique.

2D X-rays can help distinguishing defects such as a wrong die or incorrect die orientation. Such study is shown for all chips in Fig. 7.15. It can be seen that sample 3 has a totally different die which has a different orientation and size. The other

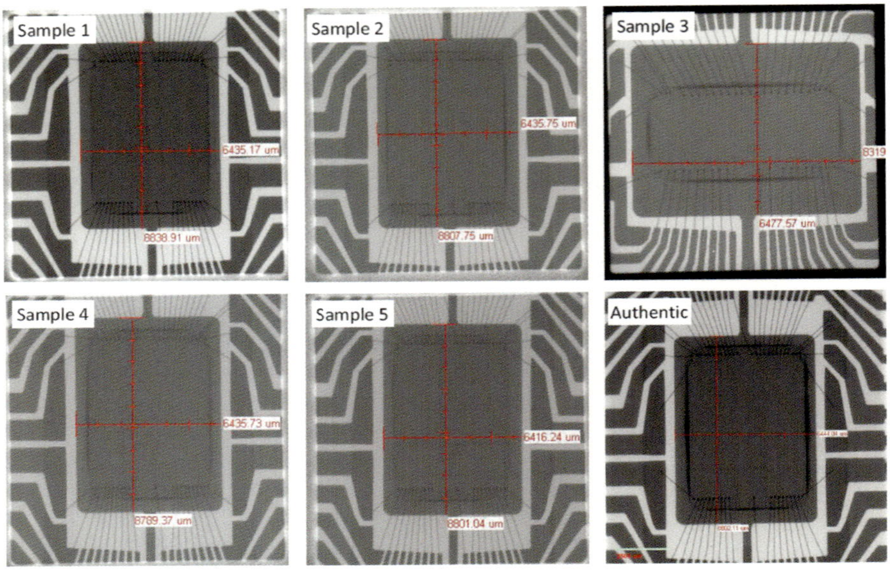

Fig. 7.15 2D X-rays of samples

four samples look very similar and even comparing to the authentic sample doesn't yield conclusive evidence though it does show a small difference in bond wire patterning. This study proves that extending the information to the third dimension is inevitable.

Figure 7.16 shows a 3D images of samples 1 and 2 where one can see signs of die face delamination on sample 1 (top left) and not in 2 (top right). Additional investigation of virtual slices show clear images of the die face delamination in all corners. (Bottom left and right). The die face delamination can be the result of recycling and is a great reliability concern. After hours of operation, such delamination can grow to the point of breaking the bond wires. Similar images revealed same results for sample 5.

The above analysis illustrates the promise of 3D X-ray Microscopy for advanced counterfeit detection, but more results need to be collected. For instance, the impact of 3D X-ray tomography on integrated circuit reliability requires further investigation. Ideally, we would like to tune the exposure to ionizing radiation such that internal defects of counterfeit parts can be seen at high enough resolution while there is also minimal impact of reliability. This shall be investigated in future work.

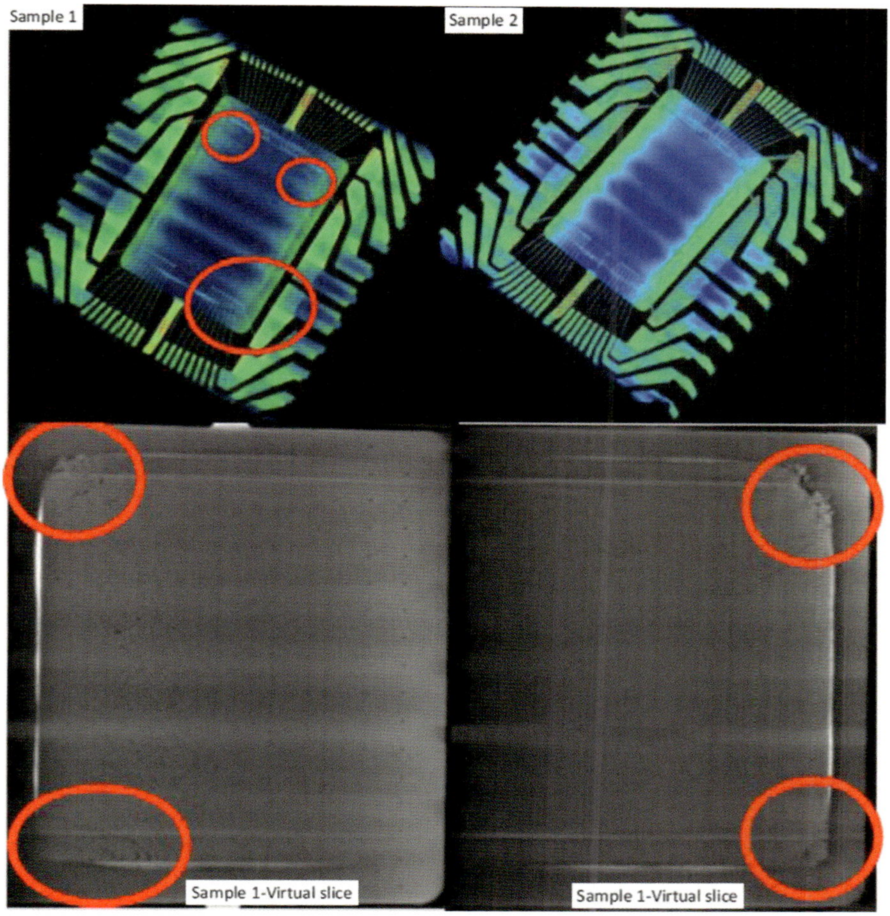

Fig. 7.16 *Top*: 3D images of the die in samples 1 and 2—showing die face delamination in sample 1. *Bottom*: virtual 2D slices of sample 1 illustrating die face delamination in all corners

7.5 Results Summary

Combining the results from the two techniques, we have been able to identify all the 5 Intel flash memory ICs as counterfeits. Qualified and quantified information are providing a complete and reliable set of data. They include both external and internal structure information of the parts. These type of information help to better decide whether a part is counterfeit or not. Results were discussed and confirmed with the SMEs at Honeywell who performed all the tests on the parts and obtained similar information.

Table 7.4 Summary of results

Observation	Instrument	Other common methods
Residue on leads	SEM/X-ray	X-ray Fluorescence (XRF), Optical microcopy
Sanding marks	SEM/EDS	Optical microscopy, XRF
Coated/filled dimples	SEM/X-ray	LSM, Optical profilometry
Dimple depth variation	SEM/X-ray	LSM, Optical profilometry
Incorrect lead plating (Sn vs. Sn/Pb)	SEM/EDS	XRF
No exposed lead base metal	SEM	XRF, Optical methods
Bent leads	SEM/X-ray	Optical methods
Metal shavings and/or tin whiskers on leads	SEM	SEM, Optical microscopy
Different die sizes	X-ray	Decapsulation, Scanning Acoustic Microscopy (SAM)
Different lead frames	X-ray	Decapsulation
Wire bond pattern variations	X-ray	Decapsulation, SAM
No barrier metal under pure Sn lead finish	SEM	XRF, Optical microscopy
Blacktopping (top and bottom surfaces)	SEM/X-ray	Destructive liquid testing

All the defects identified using the techniques described are summarized in Table 7.4. Other common techniques or instruments utilized for detection of such defects are also provided in the last column. It can be seen that using our proposed technique can successfully detect all defects nondestructively.

7.6 Summary

In this chapter, we presented two novel imaging techniques that can be applied as powerful tools for detecting defects associated with counterfeit electronic components. So far, traditional optical, digital and electron microscopy have been used to detect and characterize counterfeit ICs. With the ever-evolving techniques for counterfeiting, the counterfeit defects are now becoming more subtle and increasingly difficult to detect with prevailing techniques. Further, most physical tests utilized today provide limited information on the components under test (e.g. only 2D information might be available) and also, the tests rely heavily on the interpretation of subject-matter experts (SMEs), which brings about inconsistencies in the results. In order to alleviate these issues, two new techniques, 4D SEM and 3D X-Ray microscopy, were presented in this chapter, which are both non-destructive and do not require any sample preparation. They can give us detailed insight into the material composition and subtle exterior/interior defects of suspect ICs. These features are well beyond the capabilities of traditional 2D imaging techniques. In addition, these techniques allow us to obtain all required information in a single imaging session which can greatly reduce the time and cost of counterfeit detection procedures. The chapter also introduced a thorough quantitative approach for texture analysis, which can greatly help in interpreting data for counterfeit defects such as sanding that bring about material and texture inconsistencies. Three dimensional characterization of the dimples in IC packaging along with compositional analysis was also introduced to better detect counterfeit parts and identify further detection procedures if required. 3D X-ray analysis was introduced as an effective method to access the interior features and geometry of ICs, which would not have been possible with 2D X-Ray techniques. The chapter also showed how the identified counterfeit defects were features unique to counterfeit components, and were not found in verified authentic samples ("golden ICs"). However, one does not necessarily require a golden IC to check for the defects that were discussed in this chapter, as inconsistencies within a lot of samples can be used for counterfeit detection. Although preliminary steps of automation in counterfeit detection were introduced by providing quantitative metrics, more work needs to be done to establish a rigorous algorithm for effective automation. Future work could focus on developing such algorithms and improving the quality of the methods proposed in this chapter. Also, detailed statistical analysis is required to prove with certainty that the detection is consistent.

References

1. S. Kang, H. Cho, A projection method for reconstructing x-ray images of arbitrary cross-section. NDT & E Int. **32**(1), 9–20 (1999)
2. S. Brand, P. Czurratis, P. Hoffrogge, M. Petzold, Automated inspection and classification of flip-chip-contacts using scanning acoustic microscopy. Microelectron. Reliab. **50**(9), 1469–1473 (2010)
3. R. Tilgner, P. Alpern, J. Baumann, G. Pfannschmidt, O. Selig, Changing states of delamination between molding compound and chip surface: a challenge for scanning acoustic microscopy. IEEE Trans. Compon. Packag. Manuf. Technol. Part B: Adv. Packag. **17**(3), 442–448 (1994)
4. C. Boit, New physical techniques for ic functional analysis of on-chip devices and interconnects. Appl. Surf. Sci. **252**(1), 18–23 (2005)
5. M. Cason, R. Estrada, Application of x-ray microct for non-destructive failure analysis and package construction characterization, in *2011-18th IEEE International Symposium on the Physical and Failure Analysis of Integrated Circuits (IPFA)*, July 2011, pp. 1–6
6. T.D. Moore, D. Vanderstraeten, P.M. Forssell, Three-dimensional x-ray laminography as a tool for detection and characterization of bga package defects. IEEE Trans. Compon. Packag. Technol. **25**(2), 224–229 (2002)
7. U. Guin, D. DiMase, M. Tehranipoor, Counterfeit integrated circuits: detection, avoidance, and the challenges ahead. J. Electron. Test. **30**(1), 9–23 (2014)
8. K. Takahashi, H. Terao, Y. Tomita, Y. Yamaji, M. Hoshino, T. Sato, T. Morifuji, M. Sunohara, M. Bonkohara, Current status of research and development for three-dimensional chip stack technology. Jpn. J. Appl. Phys. **40**(4S), 3032 (2001)
9. J. Helmcke, Determination of the third dimension of objects by stereoscopy. Lab. Invest. J. Tech. Methods Pathol. **14**, 933 (1965)

10. A. Boyde, Quantitative photogrammetric analysis and qualitative stereoscopic analysis of sem images. J. Microsc. **98**(3), 452–471 (1973)
11. A. Boyde, Determination of the principal distance and the location of the perspective centre in low magnification sem photogrammetry. J. Microsc. **105**(1), 97–105 (1975)
12. G. Piazzesi, Photogrammetry with the scanning electron microscope. J. Phys. E Sci. Instrum. **6**(4), 392 (1973)
13. S. Ghosh, Photogrammetric calibration of electron microscopes. Microsc. Acta **79**(5), 419–426 (1977)
14. M.J. Roberts, K.-J.M. Söderholm, Comparison of three techniques for measuring wear of dental restorations. Acta Odontol. **47**(6), 367–374 (1989)
15. W. Hume, I. Greaves, The stereophotomicroscope in clinical dentistry. Br. Dent. J. **154**(9), 288–290 (1983)
16. J. Stampfl, S. Scherer, M. Gruber, O. Kolednik, Reconstruction of surface topographies by scanning electron microscopy for application in fracture research. Appl. Phys. A **63**(4), 341–346 (1996)
17. M. Ballerini, M. Milani, M. Costato, F. Squadrini, I. Turcu, Life science applications of focused ion beams (fib). Eur. J. Histochem. **41**, 89–90 (1997)
18. W. Drzazga, J. Paluszynski, W. Slowko, Three-dimensional characterization of microstructures in a sem. Meas. Sci. Technol. **17**(1), 28 (2006)
19. W. Beil, I. Carlsen, Surface reconstruction from stereoscopy and shape from shading in sem images. Mach. Vis. Appl. **4**(4), 271–285 (1991)
20. F. Marinello, P. Bariani, E. Savio, A. Horsewell, L. De Chiffre, Critical factors in sem 3d stereo microscopy. Meas. Sci. Technol. **19**(6), 065705 (2008)
21. D. Samak, A. Fischer, D. Rittel, 3d reconstruction and visualization of microstructure surfaces from 2d images. CIRP Ann. Manuf. Technol. **56**(1), 149–152 (2007)
22. M. Ritter, *A Landmark-Based Method for the Geometrical 3D Calibration of Scanning Microscopes*. Universitätsbibliothek (2006)
23. L. Carli, G. Genta, A. Cantatore, G. Barbato, L. De Chiffre, R. Levi, Uncertainty evaluation for three-dimensional scanning electron microscope reconstructions based on the stereo-pair technique. Meas. Sci. Technol. **22**(3), 035103 (2011)
24. T. Everhart, R. Thornley, Wide-band detector for micro-microampere low-energy electron currents. J. Sci. Instrum. **37**(7), 246 (1960)
25. O. Kolednik, The characterization of local deformation and fracture properties–a tool for advanced materials design. Adv. Eng. Mater. **8**(11), 1079–1083 (2006)
26. A. Tatschl, O. Kolednik, A new tool for the experimental characterization of micro-plasticity. Mater. Sci. Eng. A **339**(1), 265–280 (2003)
27. Mex Software from Alicona Imaging GmbH, Graz, Austria
28. Olympus Soft Imaging Solutions GmbH, Muenster, Germany
29. 3D-TOPX from SAMxPlus, Trappes, France
30. R.D. Bonetto, J.L. Ladaga, E. Ponz, Measuring surface topography by scanning electron microscopy. II. Analysis of three estimators of surface roughness in second dimension and third dimension. Microsc. Microanal. **12**(02), 178–186 (2006)
31. J.I. Goldstein, D.E. Newbury, P. Echlin, D.C. Joy, C. Fiori, E. Lifshin et al., *Scanning Electron Microscopy and X-ray Microanalysis. A Text for Biologists, Materials Scientists, and Geologists* (Plenum, New York, 1981)
32. S. Shahbazmohamadi, E.H. Jordan, Optimizing an sem-based 3d surface imaging technique for recording bond coat surface geometry in thermal barrier coatings. Meas. Sci. Technol. **23**(12), 125601 (2012)
33. M. Cason, R. Estrada, Application of x-ray microct for non-destructive failure analysis and package construction characterization, in *2011 18th IEEE International Symposium on the Physical and Failure Analysis of Integrated Circuits (IPFA)*, July 2011, pp. 1–6
34. W. Dong, P. Sullivan, K. Stout, Comprehensive study of parameters for characterizing three-dimensional surface topography I: some inherent properties of parameter variation. Wear **159**(2), 161–171 (1992)

35. W. Dong, P. Sullivan, K. Stout, Comprehensive study of parameters for characterizing three-dimensional surface topography II: statistical properties of parameter variation. Wear **167**(1), 9–21 (1993)
36. W. Dong, P. Sullivan, K. Stout, Comprehensive study of parameters for characterizing 3-D surface topography III: parameters for amplitude and some functional properties. Wear **178** (1–2), 29–43 (1994)
37. W. Dong, P. Sullivan, K. Stout, Comprehensive study of parameters for characterizing 3-D surface topography IV: parameters for characterizing spatial and hybrid properties. Wear **178**(1–2), 45–60 (1994)
38. SAE, Test methods standard; counterfeit electronic parts. Work in Progress, http://standards.sae.org/wip/as6171/

Chapter 8
Advanced Detection: Electrical Tests

Conventional counterfeit detection methods suffer from excessive test time and cost. The physical tests (see Chap. 4) require expensive equipment and, due their destructive nature, cannot be applied to all chips. Electrical tests (see Chap. 5) on the other hand require different test set ups for all the unique types of ICs one could encounter in practice (digital, analog, mixed signal, memories, processors, FPGAs, etc.). In addition, test program generation for obsolete and active ICs faces major drawbacks as well. Requiring a high-speed tester (ATE) in order to apply functional test patterns to different ICs makes it extremely expensive. It is nearly impossible to get a complete set of test vectors for an obsolete part from the original component manufacturers.

In this chapter, we will describe more advanced electrical detection of two different types of recycled ICs-field programmable gate arrays (FPGAs) and application specific integrated circuits (ASICs). Here the term advanced refers to the fact that the tests are specifically designed to target the detection of recycled counterfeit types, not the complexity of the tests themselves.

FPGAs are in the top five of counterfeited electronic components and the share of counterfeit FPGAs is expected to increase with the growth of market share of the FPGAs in the electronic industry [1]. For FPGAs, we will explain a two phase detection approach that utilizes one-class SVM classifier to classify fresh FPGAs from recycled FPGAs [2]. To detect recycled ASICs, we will describe path delay analysis [3]. The path delay information is measured during the manufacturing test process. There is no change required in current well-established design and test flows to implement this process. Statistical data analysis using Principal Component Analysis (PCA) is used to identify recycled ICs. Finally, we will also present as summary of early failure rate (EFR) analysis for detection of recycled ICs using one-class SVM.

M. Tehranipoor et al., *Counterfeit Integrated Circuits*,
DOI 10.1007/978-3-319-11824-6_8

8.1 Two Phase Detection Approach for Recycled FPGAs

While there have been some methods proposed for recycled IC detection in general, so far not much work has been done specifically on recycled FPGA detection. In this section, we will explain how recycled FPGAs exhibit anomalous behavior due to silicon aging, which can be detected using advanced electrical tests. We will first describe the aging effect on FPGAs and its implications for detection. Then, we will explain a two phase detection approach which exploits the aging characteristics of recycled FPGAs.

8.1.1 Aging and Recycled FPGAs

There are two significant characteristics of aging on FPGAs that can be exploited: (i) the performance degradation of FPGA logic and (ii) the decrease in the aging speed of FPGAs over time. Actually, all the CMOS integrated circuits have the same effects of aging, but because their reconfiguration capability, FPGAs offer greater capability to investigate the impact of aging. In other words, they do not require any additional circuitry to be added to the FPGA's base design to facilitate measurements or detection.

8.1.1.1 Aging Effect on FPGA Performance

As it is well known fact, NBTI and HCI (see Chap. 3, Sect. 3.5) significantly affect the performance of CMOS devices because of their impact on transistor threshold voltage. Even though the structure of the FPGAs is different from that of the ASIC, aging has a similar effect on FPGAs, as FPGA LUTs also show significant performance degradation over the course of their lifetime [4–6]. In [6], the authors investigated the aging effect on RO-based PUF using FPGAs. They reported that the impact of aging on FPGA LUTs running ROs is 6.7 % after 400 h of usage under high temperature and high voltage. This shift in performance provides one approach for differentiating used and fresh CMOS devices.

8.1.1.2 Decrease in Aging Speed

The other phenomenon that has been observed is that the rate of the degradation of CMOS transistors (discussed above) slows with aging. In other words, the degradation due to aging is faster when the chip is newer and lessens over time. In [6], the authors showed the aging degradation of the ROs after 200 and 400 h of aging under high temperature and high voltage. Their results show that the degradation rate after the second 200 h of aging (1.6 %) is far less than the aging degradation in the first 200 h (5.1 %), which implies that the aging speed decreases after some times of usage in the field.

In [2], we performed an experiment on two fresh FPGAs to show whether the decrease in the aging speed occurs even after a small amount of aging (prior usage). In the experiment, 224 ROs were placed on two FPGAs. Then two consecutive aging cycles were performed on both of them with the stress conditions of 125 °C and 1.8 V (nominal is 1.2 V), with each aging cycle lasting 3 h. The accelerated aging was done using a Temptronic Thermostream device TP04100A [7]. The frequency of ROs was measured in nominal conditions before and after each aging. Then the degradation rates were calculated by using following equation:

$$\Delta f_i = 100 * (\frac{f_{i,1} - f_{i,2}}{f_{i,1}}) \tag{8.1}$$

where Δf_i is the percentage degradation of i_{th} RO, and $f_{i,1}$ and $f_{i,2}$ are the frequencies before and after aging, respectively. Then, the FPGAs were left idle for a week in order to recover before starting the second aging cycle. In this way, the real decrease in the degradation speed could be observed. After waiting for a week, an average 0.257 % degradation recovery appeared, and when the second aging cycle was performed, the real degradation rate for the second cycle could be seen.

Figure 8.1 shows the distribution of the degradation rates for the first and second aging cycles for both FPGAs ring oscillators. The x-axis shows the percentage degradation rate of used and new FPGAs and y-axis shows the number of occurrences of the degraded ROs in the FPGAs. The figure clearly demonstrates that the first aging cycle exhibits much more degradation than the second one. After only

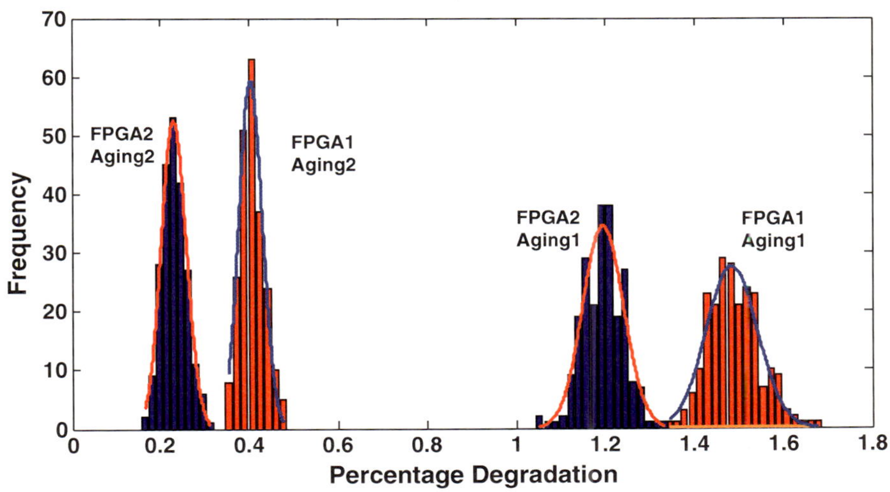

Fig. 8.1 Aging degradation distribution differences between first and second aging cycles

3 h of aging, the average aging speed went down by a factor of 3.5, thus allowing us to deduce that the aging speed is considerably different for fresh FPGAs than it is for used FPGAs. These results show that when an FPGA LUT is used for some period of time, the aging speed of the LUT decreases, so that this characteristic can be exploited to detect used (i.e., recycled) FPGAs.

That being said, since the same LUTs were aged successively, observing very low degradation rates in the second aging might be questionable. So to prove that aging speeds of used and unused FPGAs are noticeably distinct from each other, another experiment was conducted using real used FPGAs to show degradation rate difference between used and fresh FPGAs (aging speed reduction on used FPGAs). We will analyze results from this experiment to illustrate the two distinct characteristics of the used FPGAs from their fresh counterparts.

The first characteristic is that a lower degradation rate is expected from a used FPGA when the same stress conditions are applied. Figure 8.1 from the previous section shows this effect. However, this result might be biased since both aging cycles are applied to the same LUTs on the same FPGAs. Therefore, we need to see another experiment in which the same stress conditions are applied to real used FPGAs to compare their results with the results above. We conducted another experiment on two used FPGAs, which were previously aged using accelerated aging with s9234 benchmark. While used FPGA 1 was only used for 10 h, the other used FPGA was stressed for 50 h with the stress conditions of 125 °C and 1.8 V. Figure 8.2 illustrates the degradation distributions of fresh and used FPGAs. We can observe from the histograms that the fresh FPGA still undergoes far more degradation than the used FPGA even though ROs are not placed into exactly the same LUTs.

The other characteristic that differentiates used FPGAs from fresh ones is the increased variance of the degradation distribution of the ROs placed across FPGA LUTs when the same stress conditions are applied for a short period of time. It can be seen from Fig. 8.2 that the variance of the distribution of new FPGAs is smaller compared to the used FPGA degradation distributions. The reason that the new FPGAs show less variance is because the degradation differences across ROs are caused by only process variations and some environmental differences such as Vdd and temperature variations. In the case of used FPGAs, since they have already been placed under stress when they were in the field, there are different factors (change in Vth, leakage current, etc.) that affect the aging rates when the new stress is applied. Another factor is that not all parts of used FPGAs have undergone the same workload. While some parts of the FPGA might have more switching, other parts or regions might have less switching. Due to the HCI effect, the logic with more switching activity will age more. When we apply accelerated aging to an FPGA, the LUTs, which had more switching when they were in the field, will age less because they previously aged more. As a result, this will increase the variance of the degradation distribution for the used FPGA. Another factor that increases the variance of the degradation rates of ROs is that not all of the logic resources are exploited in an FPGA design. Even though logic resources generally have high utilization, there will still be some resources that are not utilized [5, 8]. Similar to the

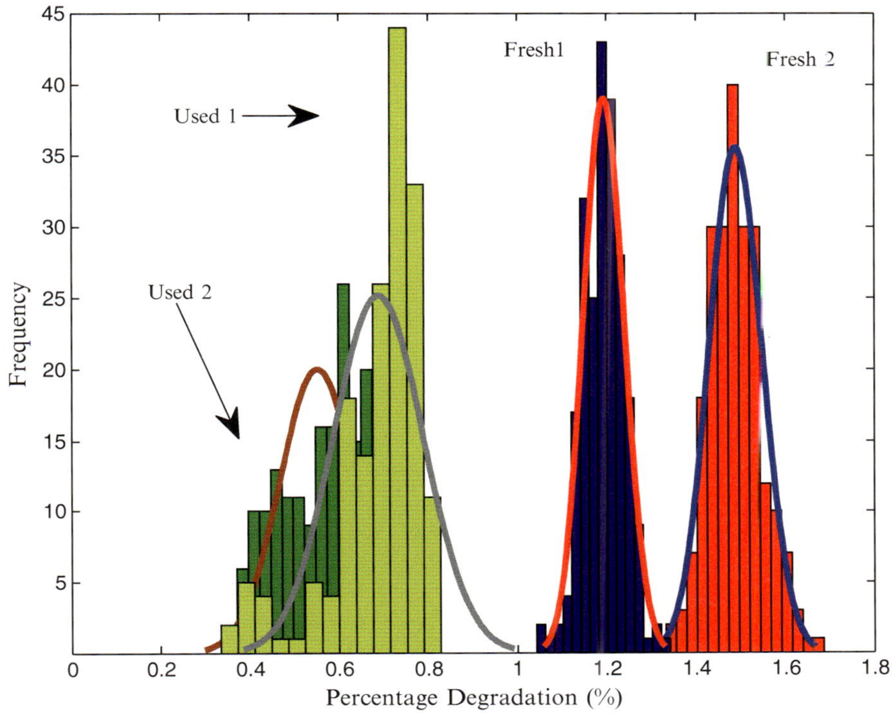

Fig. 8.2 Aging degradation distribution differences between used and fresh FPGAs

switching activity effect, unused LUTs will increase the variance of the degradation distribution of the FPGA. Figure 8.2 demonstrates that the used FPGAs present higher variance.

8.1.2 Two Phase Recycled FPGA Detection

In this section, we shall discuss a simple framework that exploits the above aging characteristics present in recycled FPGAs. Figure 8.3 illustrates the flow for the recycled FPGA detection approach in [2]. The detection approach consists of two phases. While the first phase exploits the performance degradation in the used FPGAs described in the Sect. 8.1.1.1, the second phase takes advantage of the decrease in degradation speed present in the used FPGAs addressed in the Sect. 8.1.1.2. Note that in both phases, it is assumed that there exists golden (known fresh) FPGAs. A one-class SVM classifier is trained on the information obtained from these fresh FPGAs. One-class SVM and each phase will be briefly explained in the next subsections.

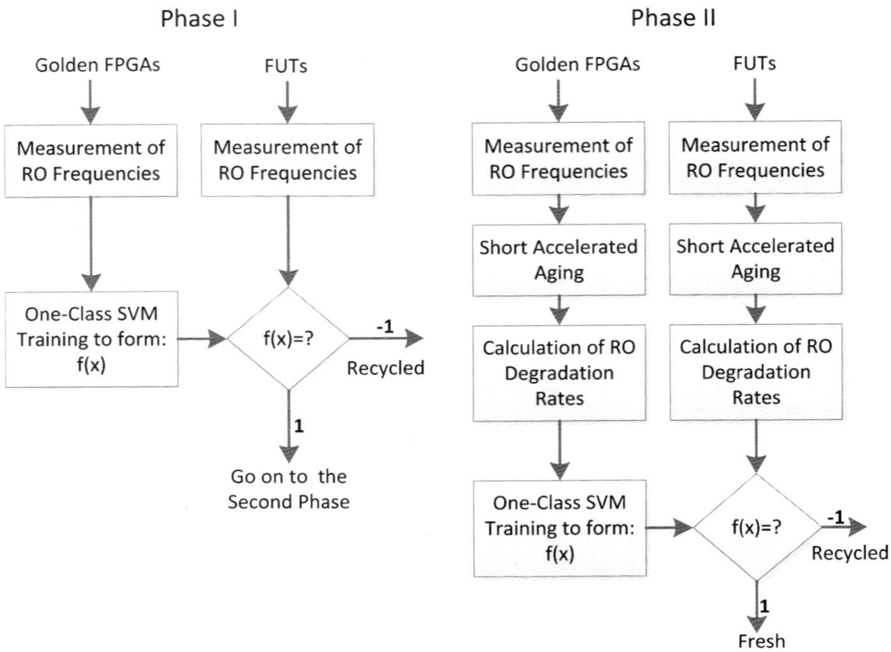

Fig. 8.3 Proposed recycled FPGA detection flow

8.1.2.1 One-Class SVM

Support Vector Machines (SVM) are generally employed for binary classification problems where the SVM is trained using sample feature data from two object classes in a training dataset. With the given dataset, SVM creates a decision function, which takes as input a feature vector from an unknown object and outputs its predicted class. To use classical SVM, data is needed from both classes, but such data may not be always available. For example in our case, we may be able to obtain frequency distributions and degradation of fresh FPGAs as our first class. However, we do not have a second class because we don't have prior knowledge about the given devices (i.e., recycled FPGAs we might encounter might be aged with different workloads and for different amounts of time). For such problems, there exist one-class classification algorithms.

One-class SVM was first introduced in [4] by Scholkopf et al. One-class SVM creates a function f which takes $+1$ in a small region, which is formed by using the training samples, and -1 elsewhere. Generally a kernel function is employed in one-class SVM to map the data points into a feature space $\mathbf{H}$, and then the feature vectors are separated from the origin with maximum margin. The function can be expressed by the following expression:

$$f(\mathbf{x}) = \begin{cases} +1 & \text{if } x \in \mathbf{H} \\ -1 & \text{if } x \in \bar{\mathbf{H}} \end{cases} \tag{8.2}$$

where $\mathbf{x}$ is the feature vector and $\mathbf{H}$ is the feature space. Let the training dataset be $\mathbf{X} = [\mathbf{x}_1, \ldots, \mathbf{x}_n]$ where $\mathbf{x}_1, \ldots, \mathbf{x}_n$ are the feature vectors and $\mathbf{X}$ is the total training samples. Let Φ be the feature map $\mathbf{X} \rightarrow \mathbf{H}$ which transforms the training points into another space using the kernel function. Then, data points are needed to be separated from the origin, hence the following quadratic programming problem needs a solution.

$$\min_{w,\phi,\xi_i} \frac{1}{2}\|w\|^2 + \frac{1}{\nu m}\sum_{i=1}^{m} \xi_i - \rho \tag{8.3}$$

subject to $(w \cdot \Phi(\mathbf{x}_i)) \geq \rho - \xi_i \quad i = 1, 2, \ldots, m \quad \xi_i \geq 0$ where ξ_i is the slack variable and ν is used to characterize the solution by setting an upper bound on the training samples which are classified as outlier and setting a lower bound on the number of support vectors.

Using the Lagrange multiplier and a kernel function to calculate dot product gives us the following decision function.

$$\begin{aligned} f(\mathbf{x}) &= sgn(w \cdot \Phi(\mathbf{x}_i) - \rho) \\ &= sgn(\sum_{i=1}^{m} \alpha_i K(\mathbf{x}, \mathbf{x}_i) - \rho) \end{aligned} \tag{8.4}$$

where α_i is the ith Lagrange mutliplier and ρ and w are used to create a hyperplane which separates all the data points from the origin.

There are different kernel functions to be used with the SVM algorithm such as linear, polynomial, and radial basis function (RBF) kernels. In this approach, the following RBF kernel is employed:

$$K(\mathbf{x}, \mathbf{x}_i) = \exp(-\gamma|\mathbf{x} - \mathbf{x}_i|^2\gamma) \tag{8.5}$$

In Eq. (8.5), the parameter γ defines how far a single training example can have effect. When the γ has small value the RBF kernel has a wide boundary containing more training examples. On the other hand, if the γ has large value, the RBF kernel contains less training examples in it.

8.1.2.2 Phase 1

Figure 8.3 (left) displays the first detection phase of the recycled FPGA detection approach. The first phase is used for the easy-to-detect recycled FPGAs, thereby bypassing the second phase which involves aging the FPGAs. Phase I is rather straightforward, fast, and low-cost. The main idea of this phase is to exploit the performance degradation of the used FPGAs over time. This phase involves measurement of RO frequencies to obtain the performance distribution of the fresh (golden) FPGAs and testing the FPGA under test (FUT) with the golden FPGA data using one-class SVM. This phase starts with placing n ring oscillators (ROs) and measuring their frequencies for each of the m golden FPGAs. Then, the second step is to create a decision function using the one-class SVM by training it with the RO frequencies. The training data that is used as follows:

$$\begin{aligned} \mathbf{F} &= [\mathbf{f}_1, \mathbf{f}_2, \ldots, \mathbf{f}_m]; \\ \mathbf{f}_i &= [f_1, f_2, \ldots, f_n] \end{aligned} \tag{8.6}$$

In Eq. (8.6), **F** is the total training dataset, and $\mathbf{f}_1, \mathbf{f}_2, \ldots, \mathbf{f}_m$ are the feature vectors for m FPGAs. Each feature vector in **F** includes the n RO frequencies as features. The frequency models for the frequencies in the feature vectors are different for fresh FPGAs and used FPGAs. For the fresh FPGAs, each frequency information in $\mathbf{f}_i$ contains the following three components; f_{nom}, $f_{intra,i,j}$ and $f_{inter,i}$. f_{nom} is the nominal frequency which is constant for every RO in every FPGA. $f_{inter,i}$ and $f_{intra,i,j}$ are the deviations induced by inter- and intra-die manufacturing variations. The sum of these components forms the frequency information for one RO. For the used FPGAs, there is one more component which affects the frequency of the RO which is the aging effect $\Delta f_{aging,i,j}$. Because of the prior usage of the device there will be an aging effect which decreases the frequency of ROs in used FPGAs.

With the given training dataset, the one-class SVM is trained and a decision boundary is formed to classify the FUTs. Some FUTs with gross degradation in the field can be detected by Phase I. Some FPGAs with more subtle degradation however may not be detected in Phase I. These different cases are illustrated in Fig. 8.4. In the figure, dots represent the frequencies of the FPGAs before their usage and arrows represent their frequency change after their usage in the field. The red arrows and black arrows highlight the cases that are more easily detectable and less easily detectable by Phase I respectively. Essentially, only those FUTs with ROs that pass the threshold (determined by one-class SVM decision boundary) are detectable by Phase I. The cases illustrated in the figure are summarized as follows:

1. **FPGA starting in slow corner:** If an FPGA has low RO frequencies when it is new (slow FPGA), it can go outside the golden distribution after short period of usage. Such an FPGA could be easily detected by Phase I.
2. **FPGA starting near nominal:** If an FPGA has RO frequencies in the center region of the golden distribution, it needs more usage time to be detected by Phase I.
3. **FPGA starting in fast corner:** If an FPGA has RO frequencies at the fast end of the distribution when new (fast FPGA), it needs far more aging to be detected in Phase I.

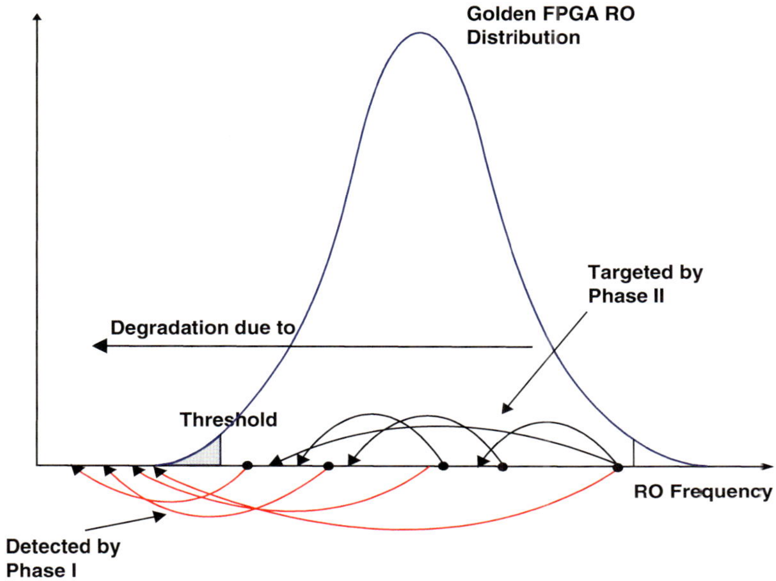

Fig. 8.4 Phase I and phase II detection regions

Table 8.1 Usage information of FUTs

Usage	50 h	10 h	6 h	Fresh
# of FUTs	4	8	4	4

For the FPGAs in the nominal and fast corners that do not have enough aging to be detected in Phase I (i.e., stay above the threshold/boundary decided by one-class SVM), the second phase is developed. Phase II does not rely on raw frequencies like Phase I and instead exploits the difference in aging rate between new and recycled FPGAs (see Sect. 8.1.1.2). Phase II does not depend on the starting corner of the FPGA and therefore should be able to detect the cases not covered by Phase I (shown with black arrows in Fig. 8.4).

Experimental results shows that phase I is effective for gross outliers and confirms the cases mentioned above. The one-class SVM is employed to form a decision function using the training samples discussed above with 20 fresh FPGAs. 224 RO frequencies were used as feature vector for each FPGA. In this phase, to obtain better results we need large golden samples, hence using 224 ROs as features takes time in terms of training. Therefore, Principal Component Analysis (PCA) was deployed in order to reduce the dimension of the feature vectors (224 features) to 13 (according to Kaiser's rule for selecting the principal components [9]). SVM was fed with 20 test data shown in Table 8.1. Accelerated aging was performed on 16 FPGAs: 50 h on 4 FPGAs, 10 h on 8 FPGAs, and 6 h on 4 FPGAs. The SVM classified the four fresh FPGAs correctly, and it also classified 4 FPGAs as recycled. The four detected FPGAs are the ones used for 10 h. The reason, as mentioned earlier and showed in Fig. 8.4, is because these FPGAs have lower RO frequencies and they can go outside the golden distribution with less usage. As the results suggest, this phase can be effective if the training set is large enough to cover as much fresh FPGA data as possible.

8.1.2.3 Phase II

The steps of the second phase of the two phase detection method are shown in Fig. 8.3 (righthand side). In short, it involves aging golden (known fresh) and unknown FUTs and distinguishing them based on the rate of degradation. It begins by analyzing behavior of m fresh (golden) FPGAs. First, n ROs are placed on the fresh FPGAs and their initial frequencies are measured in a controlled environment (under nominal voltage Vdd_{nom} and temperature T_{nom}). Next, accelerated aging is performed on the fresh FPGAs. This is accomplished by running the n ROs (note that any other circuitry could be implemented on FPGAs as well) while stressing the FPGAs with high voltage Vdd_{ref} and high temperature T_{ref} for Δt time. After aging, the n RO frequencies are re-measured at nominal conditions and their percentage degradation is computed using Eq. (8.1). The one-class SVM classifier is trained using the n ROs from the m golden samples. The training data is formulated as follows:

$$\mathbf{x}_i = [\Delta f_1, \Delta f_2, \ldots, \Delta f_n] \tag{8.7}$$

In Eq. (8.7) $\mathbf{x}_i$ denotes the i_{th} FPGA, and it includes n RO degradation rates as its features. So the total training set can be denoted by:

$$\mathbf{X} = [\mathbf{x}_1, \mathbf{x}_2, \ldots, \mathbf{x}_m] \tag{8.8}$$

where we have m number of training samples. Based on experimental results in [2], if an FPGA is used long enough for a period of time, the degradation rates for used and fresh FPGAs have little if any overlap, so not many training samples are needed. Then the initial measurement, aging, second measurement, and percentage degradation calculation steps are repeated to obtain n features for each FPGA under test (FUT). Finally, an SVM classifier $f(\mathbf{x})$ is generated based on the training data and then used to classify each FUT as fresh or recycled.

Figure 8.5 shows the result of the one-class SVM using 20 golden FPGAs for training and 20 FUTs (the same FUTs shown in Table 8.1). The samples shown with "▷" markers are the test data and the samples with "green circle" markers are the training data. 20 FUTs are shown in the Table 8.1 and it contains 4 fresh FPGA data to test the effectiveness of the method. Sixteen FUT data contains data from FPGAs used for 6, 10 and 50 h using s9234 benchmark at the conditions of high temperature and high voltage. The figure clearly indicates that the one-class SVM can classify the fresh FUTs correctly. The remaining 16 used FPGAs were detected as outliers

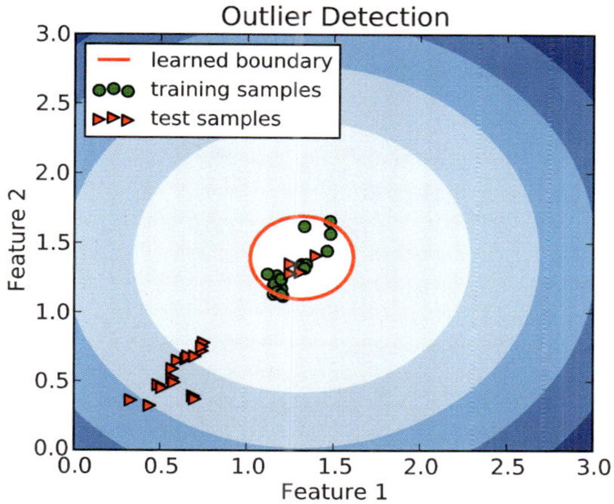

Fig. 8.5 One-class SVM boundary and outlier detection

even though some of them were only aged for 6 h using a very small benchmark. These results show the effectiveness of using degradation rates and one-class SVM to differentiate the recycled FPGAs.

Note that one drawback to this second phase is that the FUTs need to be aged, thereby impacting the FUTs. The reported average performance degradation of the above 20 new FPGAs after applying accelerated aging for 3 h is 1.283 % [2]. The effect of this drawback can be reduced by decreasing the aging time. Additionally, we do not need to apply the second phase to every FUT, but to a random sampling of FUTs in a batch. If any of them are classified as recycled with high confidence, we can discard the entire batch.

8.2 Path-Delay Analysis

Path delay fingerprinting [3] was proposed to screen recycled ICs without adding extra hardware in the design. Since these recycled ICs have been used in the field, the performance of such ICs must have been degraded due to the impact of aging. Due to the process variation, the delay distribution of the paths lies within the specified range. The fingerprint of the new ICs can be generated during manufacturing test and stored in a secured database. Due to negative/positive bias temperature instability (NBTI/PBTI) and hot carrier injection (HCI), the path delays in recycled ICs will become larger. The larger path delays indicate that the higher probability of being an IC used for a long period of time in the field. In path delay fingerprinting approach, statistical data analysis is used to classify recycled (aging causes the delay variation) and new ICs (process variation causes the delay variation). Since the path delay information is measured during the manufacturing test process, no extra hardware circuitry is required for this technique. Note that no change is required in current well-established design and test flows.

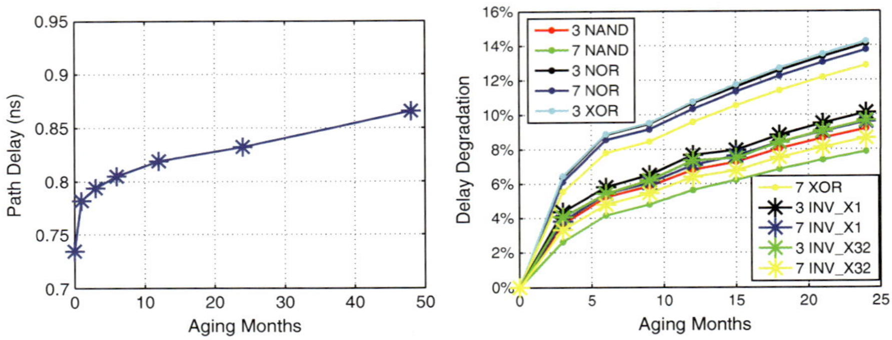

Fig. 8.6 Path delay degradation due to aging [3]. (**a**) Delay of an arbitrary path and (**b**) Delay degradation of different gate chains

8.2.1 *Impact of Aging on Path Delays*

Aging causes a non-recoverable shift of device parameters over time. As we explained in Chap. 3, Sect. 3.5, NBTI and HCI are the two major phenomena that cause these parametric shift. NBTI increases the absolute value of the PMOS threshold voltage, which results in increasing gate delay [10, 11]. HCI occurs in NMOS devices caused by the trapped interface charge at Si/SiO_2 surface near the drain end during switching that results a non-recoverable V_{th} degradation [10, 12]. Since recovered ICs have been aged, the path delay of recycled ICs will be increased.

Figure 8.6 shows the delay degradation of a randomly selected critical path of ISCAS'89 benchmark circuit s38417 when the circuit was driven with a random workload (random functional patterns are applied to the primary input). The path was aged for 4 years, using simulation, with NBTI and HCI effects at room temperature. We can observe from Fig. 8.6a that the degradation of the path used for 1 year is around 10 % while if the circuit is used for 4 years, the degradation is about 17 %, indicating that most aging occurred at the early usage phase of the circuit. Figure 8.6b presents the delay degradation of different chains, consisting of INVX1, INVX32, AND, NOR, and XOR gates, after 2 years of aging. We can see that different chains age at slightly different rates, which depends on the structure of the gates. The XOR gate chain has the highest aging rate which will help to select the paths for fingerprinting.

8.2.2 *Path Delay Fingerprinting*

Figure 8.7 shows the flow for identifying recycled ICs using path delay fingerprints and statistical analysis. It consists of three major steps:

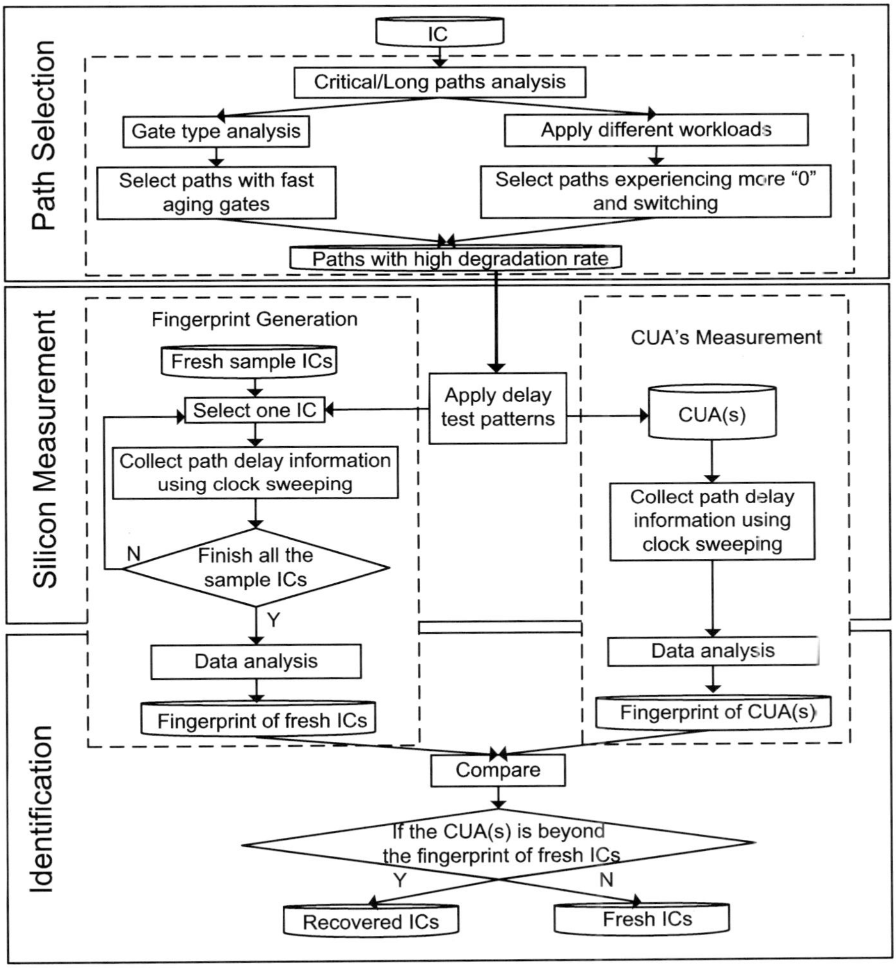

Fig. 8.7 Recycled IC identification flow [3]

1. Path Selection: In this stage, paths are selected for generating the fingerprints. Due to the large number of critical and long paths in a circuit, the paths with higher degradation rate are selected. We select paths based on two criteria: (i) those paths having large number of XOR gates and (ii) those with more number of gates having "0" at their inputs during a random workload. We prefer the paths with higher degradation rates for fingerprint generation.
2. Silicon Measurement: Using clock sweeping technique (see below in Sect. 8.2.3), the delay information of these paths are measured either during manufacturing tests or during authentication on a large sample of new ICs. In this stage, we measure the delays for the same paths, registered during manufacturing tests of

circuits under authentication (CUAs). The measurement environment should be controlled properly to keep the temperature stable as it may also impact the delay of a path significantly.

3. Identification: Once the path delays in all the new ICs are measured, statistical data analysis will be used to generate the fingerprint. Two statistical data analysis methods can be used: simple outlier analysis (SOA), or principal component analysis (PCA). If the CUA is outside of the convex created by the new ICs (convex hull), there is a high probability that the CUA is recycled.

Additional details on the above steps are discussed in the subsections below.

8.2.3 Clock Sweeping

Clock sweeping technique was introduced in [13] for identifying recycled ICs. This technique does not require any area overhead as it utilizes the common design or test processes. It uses path delay information to create unique binary identifiers. This technique represents a novel improvement on existing ideas for several reasons. First, this technique can be applied to ICs already in the supply chain, including legacy designs. Second, it uses data that can be obtained through use of existing pattern sets and testing hardware capabilities. Finally, no additional hardware is necessary—there is no area, power, or timing overhead to the technique.

Clock sweeping is the process of applying patterns to a path multiple times with different frequencies to find a frequency at which the path cannot propagate its signal, often for purposes of speed binning. By observing the frequencies at which the path can and cannot propagate its signal, we can measure the delay of the path with some degree of precision. Our ability to perform clock sweeping on a path is limited by the degree of control we have on the clock that controls the capturing memory elements (i.e., the flip-flops), the degree to which we can excite paths in the circuit, and the lengths of the paths in the IC.

Figure 8.8 shows a visual example of clock sweeping being performed on several paths. Assume that paths P1 through P8 are paths in the circuit which end with a capturing flip-flop, and have some delay in nanoseconds. Each of the eight paths can be swept (tested) at the frequencies f_1 through f_5. All paths are able to propagate their signal at f_1, as this is the rated frequency of the IC design. However, at f_2, the path P3 will usually fail to propagate its signal. At frequency f_3, path P3 will always fail to propagate its signal. Path P8 will succeed in propagating its signal at all five clock frequencies in this example, because it is too short to test with clock sweeping. All of the paths have some number of frequencies they will pass at, some they may fail at, and some they are guaranteed to fail at. Process variations change which frequency each path will fail at between different ICs.

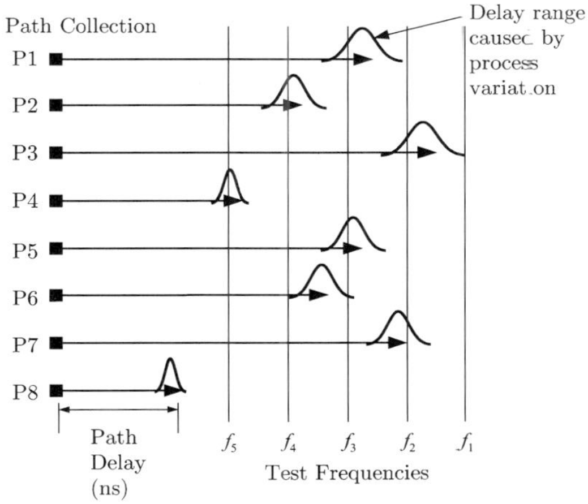

Fig. 8.8 Clock sweeping [13]

8.2.4 Data Analysis

The dimension of the collected data is large, several hundreds, even though we collect a small percent of long, or critical paths. The delay information for each path represents one dimension. It is thus necessary to reduce the dimensions to create the fingerprint. Principle Component Analysis (PCA), one of the popular multivariate analysis methods, is used to reduce this large data dimensions [14]. PCA uses an orthogonal transformation to convert a set of linear correlated variables into a much smaller set of linearly uncorrelated variables. These smaller number of uncorrelated variables are called principal components. We use an inbuilt MATLAB function to compute the principle components [15]. Due to the scope of this book, we are not going to describe PCA in detail. The interested readers can find more information about PCA in [14].

After computing the principle components, a 3D convex hull is plotted to visually show the fingerprint. A convex hull represents the smallest convex region enclosing a set of points in n-dimensions space. In this technique, we use 3D convex hull plot using inbuilt MATLAB function [16].

8.2.5 Results

Figure 8.9 shows the PCA results of ICs having inter-die and intra-die variation of v_{th}, L, and T_{ox} variation of 8, 8, 2 % and 7, 7, 2 % respectively. The detection rate of recycled ICs having aged 6 months, and 1 year are 99.3, and 100 %, respectively. Figure 8.9a, b show the new ICs' fingerprint (the convex hull) and the recovered ICs used for 6 months and 1 year, respectively. The recovered ICs used for longer times

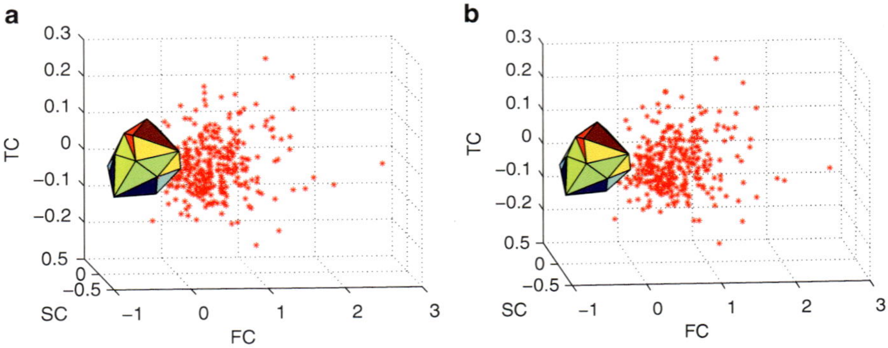

Fig. 8.9 PCA results of ICs [3]. (**a**) PCA results for ICs with 6 months of aging, and (**b**) PCA results for ICs with 1 year of aging

are easier to detect, as seen in these figures. The detection rate reduces significantly when the ICs used shorter period of time. For example, the rate reduced to 72.7 % when the ICs are used only for 1 month.

8.3 Early Failure Rate (EFR) Analysis

Another approach using Support Vector Machines (SVMs) have been proposed in [17] to detect recycled ICs. The authors train a one-class SVM using parametric measurements of new ICs. The validation of the model is performed by ICs through product reliability op-life tests. No additional cost is required to collect the data as most of the ICs require early failure rate (EFR) analysis by using burn-in tests at an elevated temperature and voltage, to reduce failures in the field. The model works as follows. First initial parametric measurements, such as V_{min}, F_{max}, and I_{ddq}, data are collected from a trustworthy manufacturer to train a one-class classifier. Then the same parametric measurements are performed to ICs under authentication and submitted to the model to classify ICs that belong to recycled type.

8.4 Summary

In this chapter, we have discussed several techniques to detect recycled ICs. We have presented an effective recycled FPGA detection method by exploiting the performance degradation and the aging speed slowdown of the used FPGAs. The proposed method consists of two phases and both phases rely on machine learning via support vector machines (SVM) for classification. The results from Xilinx FPGAs showed that the second phase of the proposed method is very powerful for detecting the used FPGAs but it requires a short amount of accelerated aging. The first phase was shown to be effective to identify the gross outliers without the need for accelerated aging making it very low-cost and easy to implement. We have described path delay analysis to detect recycled ASICs as well. This technique uses Principal Component Analysis as a statistical data analysis tool to classify recycled ICs from the authentic ones. The inherent advantage of this technique is that it does not require any modification to the well-established design and test flow.

There are several challenges and limitations that must be overcome to make these techniques successful. These approaches have exploited the aging phenomenon to detect recycled ICs. These approaches require that the performance measurements of new authentic (golden) ICs be collected and analyzed. This represents a major challenge for legacy parts when authentic ICs may not be available. Furthermore, large process variations in lower technology nodes can make it very difficult to separate recycled ICs from a batch when the variation from process variation outpaces aging degradation. To address these limitations, we will present different design-for-anti-counterfeit (DFAC) measures in the following chapters for the effective detection and avoidance of counterfeit ICs.

References

1. IHS iSuppli, Top 5 Most Counterfeited Parts Represent a $169 Billion Potential Challenge for Global Semiconductor Market (2011)
2. H. Dogan, D. Forte, M. Tehranipoor, Aging analysis for recycled fpga detection, in *2014 IEEE International Symposium on Defect and Fault Tolerance in VLSI and Nanotechnology Systems (DFT)* (IEEE, 2014), pp. 171–176
3. X. Zhang, K. Xiao, M. Tehranipoor, Path-delay fingerprinting for identification of recovered ics, in *Proc. International Symposium on Fault and Defect Tolerance in VLSI Systems*, October 2012
4. B. Schölkopf, J.S.-T.R.C. Williamson, A.J. Smola, J.C. Platt, Support vector method for novelty detection. NIPS **12**, 582–588 (1999)
5. S. Srinivasan, P. Mangalagiri, Y. Xie, N. Vijaykrishnan, K. Sarpatwari, Flaw: Fpga lifetime awareness, in *Proceedings of the 43rd annual Design Automation Conference* (ACM, 2006), pp. 630–635
6. A. Maiti, L. McDougal, P. Schaumont, The impact of aging on an fpga-based physical unclonable function, in *2011 International Conference on Field Programmable Logic and Applications (FPL)* (IEEE, 2011), pp. 151–156
7. Temptronic ThermoStream: TP04100A [Online], Available: http://www.artisantg.com/info/temptronic_tp04100a_thermostream_applica-tion_manual.pdf
8. T. Tuan, B. Lai, Leakage power analysis of a 90nm fpga, in *Proceedings of the IEEE 2003 Custom Integrated Circuits Conference, 2003* (IEEE, 2003), pp. 57–60
9. H.F. Kaiser, The application of electronic computers to factor analysis. Educ. Psychol. Meas. **20**, 141–151 (1960)
10. S. Mahapatra, D. Saha, D. Varghese, P. Kumar, On the generation and recovery of interface traps in MOSFETs subjected to NBTI, FN, and HCI stress IEEE Trans. Electron Dev. **53**(7), 1583–1592 (2006)

11. K. Uwasawa, T. Yamamoto, T. Mogami, A new degradation mode of scaled p+ polysilicon gate pMOSFETs induced by bias temperature (BT) instability, in *International Electron Devices Meeting, 1995 (IEDM '95)*, Dec 1995, pp. 871–874
12. P. Heremans, R. Bellens, G. Groeseneken, H. Maes, Consistent model for the hot-carrier degradation in n-channel and p-channel MOSFETs. IEEE Trans. Electron Dev. **35**(12), 2194–2209 (1988)
13. N. Tuzzio, K. Xiao, X. Zhang, M. Tehranipoor, A zero-overhead ic identification technique using clock sweeping and path delay analysis, in *Proceedings of the Great Lakes Symposium on VLSI, ser. GLSVLSI '12* (ACM, New York, 2012), pp. 95–98. [Online], Available: http://doi.acm.org/10.1145/2206781.2206806
14. S. Wold, K. Esbensen, P. Geladi, Principal component analysis. Chemom. Intell. Lab. Syst. **2**(1), 37–52 (1987)
15. MathWorks, Principal component analysis of raw data, http://www.mathworks.com/help/stats/pca.html
16. MathWorks, Convex Hulls, http://www.mathworks.com/help/matlab/math/convex-hulls.html#bsp2xgl
17. K. Huang, J. Carulli, Y. Makris, Parametric counterfeit IC detection via support vector machines, in *Proc. International Symposium on Fault and Defect Tolerance in VLSI Systems* (2012), pp. 7–12

Chapter 9
Combating Die and IC Recycling

In today's electronic component supply chain recycled and remarked parts account for a significant percentage of counterfeit components. The detection of these parts poses a significant challenge to the global electronic component supply chain due to the lack of efficient, robust, and low-cost detection and avoidance technologies. While there are electrical and physical tests defined in the standards [1–3] to identify counterfeit ICs, these approaches are usually characterized by excessive test time, high cost, and low confidence [4–8]. In this chapter, we discuss alternative approaches that can be integrated into new components at very low costs and can enable fast detection of recycled ICs. These approaches are a part of the design methodology to deter counterfeit components, which we term as *design-for-anti-counterfeit (DFAC)* measures.

Several approaches, discussed in Chap. 8, have been proposed to detect recycled ICs. According to path-delay fingerprinting [9], used components can be differentiated from their genuine counterparts as their path delay distribution changes (described in Chap. 8). This technique, however, presents several shortcomings, one of which is that it requires data from genuine ICs and cannot be easily applied to analog/RF/mixed-signal devices. In [10], a statistical approach was presented to distinguish recycled ICs by measuring electrical parameters and using a one-class support vector machine (SVM). Like path-delay fingerprinting, this technique requires a large number of genuine samples for SVM training. This may not be feasible as there are thousands of different types of components available in the supply chain, making it difficult to find large numbers of genuine samples. Thus, it is of utmost importance that we develop a new, practical DFAC structures that will enable easy counterfeit detection without the need for existing expensive test methods and/or genuine ICs.

Figure 9.1 illustrates the DFAC technologies we have identified for recycled and remarked counterfeit avoidance. The x-axis and y-axis represent the counterfeit types and component types respectively. The component types on the y-axis are

M. Tehranipoor et al., *Counterfeit Integrated Circuits*,
DOI 10.1007/978-3-319-11824-6_9

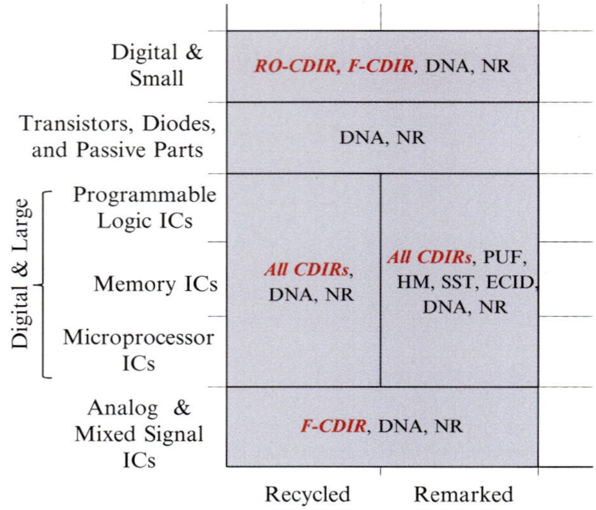

Fig. 9.1 Counterfeit avoidance technologies

arranged top to bottom from lowest to highest frequency of counterfeit incidents in the supply chain [11]. The technologies in red, represent the solutions to be presented in the latter sections of this chapter.

To track ICs throughout the supply chain, each IC needs to be tagged with a unique ID. This electronic chip ID (ECID) can be easily read during the chip's lifetime. The conventional approach for writing the unique ID into a non-programmable memory (such as One-Time-Programmable [OTP], ROM, etc.) requires post-fabrication external programming, such as laser fuses [12] or electrical fuses (eFuses) [13]. The eFuse is gaining popularity over the laser fuse because of its small area and scalability [13]. Alongside ECID, silicon physical unclonable functions (PUFs) have received much attention from the hardware security and cryptography communities as a new approach for IC identification, authentication, and on-chip key generation [14–18]. Silicon PUFs exploit inherent physical variations (process variations) that exist in modern integrated circuits. These variations are uncontrollable and unpredictable, making PUFs suitable for IC identification and authentication [19, 20]. The variations can help generate a unique signature for each IC in a challenge-response form, which allows later identification of genuine ICs.

Similar to PUFs, hardware metering (HM) can be applied to detect new remarked ICs. These metering approaches can be either passive or active. Passive approaches uniquely identify each IC and register the IC using challenge-response pairs. Later, suspect ICs taken from the market are checked for proper registration [15, 17, 21–24]. Active metering approaches, however, lock each IC until it is unlocked by the IP holder [20, 25–29]. This locking can be done in a variety of ways, which include the following: (1) initializing ICs to a locked state on power-up [20]; (2) employing combinational locking by, for instance, scattering XOR gates randomly throughout the design [27–29]; and (3) adding a finite-state machine

(FSM) which is initially locked and can be unlocked only with the correct sequence of primary inputs [26, 30]. Along with hardware metering, secure split test (SST) [31] is proposed to detect new remarked ICs.

A large portion of the supply chain is populated by active and obsolete components. There is no opportunity for adding any extra hardware to create a chip ID in those designs. For tagging such active and obsolete components, we need to create package IDs that do not require access to designs. No package modifications are allowed during the generation of package IDs (see Chap. 12). These IDs can be used for new components as well. At present, only DNA markings (DNA) [32] are commercially available for that purpose. Detection of counterfeit parts can be accomplished via detailed or fast authentication. However, detailed DNA validation is extremely time-consuming and costly [33] which makes it impractical for detecting recycled parts. If the counterfeiters add the same DNA or a different mechanism to the chip after recycling that shines the same light, this technology will be ineffective for fast-authentication (only observing color) of remarked ICs. Nanorods (NR) technology [34], not yet commercially available, may also suffer from similar issues.

The technologies discussed so far (ECID, PUFs, HM, and SST), unfortunately, are not suitable for detecting recycled ICs as long as the counterfeiters maintain the same grade (e.g., commercial grade component remains same). In addition, many of these technologies cannot be implemented on small parts because of their large area overhead. They are also inapplicable on analog and mixed-signal components due to the difference in technologies. DNA and NR have their own challenges for use in IC authentication. In this chapter, we present very low-cost structures that can be implemented in the full spectrum of components to detect recycled and remarked types. These technologies are added to the die, making them suitable for new components.

The above discussion highlights the major challenges that must be overcome in order to realize more effective DFAC measures. In this chapter, we address the shortcomings of prior work by (1) developing separate measures for analog and digital components as they are of different sizes and use different manufacturing technologies; (2) keeping the cost/overheads of adding the DFAC measures as low as possible; and (3) enabling fast authentication with low-cost test devices that do not require genuine ICs for the purpose of comparison. We meet these objectives by presenting several new combating die and IC recycling (CDIR) structures. In Sect. 9.1, we introduce two lightweight ring-oscillator-based CDIR structures suitable for both large and small digital ICs. We call these structures as simple RO-CDIR [35] and NBTI-aware RO-CDIR [36]. The second NBTI-aware RO-CDIR exploits aging much better than the simple RO-CDIR so that it is able to capture very short usage time for a chip. In Sect. 9.2, we describe two different structures based on anti-fuse that can measure the life of large digital ICs with tunable accuracy and cost. In Sect. 9.3, we present two fuse-based CDIR (F-CDIR) structures primarily aimed at analog and small ICs. A very low-cost measurement device such as a multimeter can authenticate the component with these F-CDIRs.

Depending on the size of the chip and the accuracy required in measuring the IC usage, one can select one or a combination of these CDIR structures for recycled IC detection. Note that in this chapter, we only address the remarking of recycled ICs, not the remarking of new ICs.

9.1 RO-Based CDIR Sensor

The first set of avoidance measures are to be taken by placing ring-oscillator-based CDIRs (RO-CDIRs) in the digital ICs. This simple elegant structure utilizes aging efficiently to authenticate ICs as counterfeit or not. In the following, we will describe aging phenomenon in detail, and then present two different versions of RO-CDIR.

Recycled ICs are characterized by aging, i.e., prior usage has taken its toll on the components' life and performance. A shift in the components' parameters due to aging will occur when they are used in the field for some time, which leads to the development of parametric defects and anomalies in the component. Aging of a component used in the field can be attributed to two major, distinct phenomena (which are becoming more prevalent as the technology scales down). They are negative-bias temperature instability (NBTI) and hot carrier injection (HCI) which are prominent in PMOS and NMOS devices, respectively. The detailed description of these two have been discussed in Chap. 3 Sect. 3.5. NBTI occurs in p-channel MOS devices stressed with negative gate voltages and elevated temperatures due to the generation of interface traps at the $Si - SiO_2$ interface. Removal of the stress can anneal some of the interface traps, but not completely. As a result, it manifests as the increase in threshold voltage (V_{th}) and absolute off current (I_{off}) and the decrease of absolute drain current (I_{DSat}) and transconductance (g_m). HCI occurs in NMOS devices caused by the trapped interface charge at $Si - SiO_2$ surface near the drain end during switching. It results in non-recoverable V_{th} degradation. These two aging mechanisms lead to the increased delay in the components' internal paths, which ultimately reduces the component's operating speed. Now the obvious question is *can aging help us to detect recycled ICs?* And, the answer is *yes*!

Prior approaches [9, 10] for the detection of recycled ICs, have exploited this aging phenomenon. These approaches require that the performance measurements of fresh chips be collected and analyzed, a challenge for legacy parts when golden ICs may not be available. Furthermore, large process variations in lower technology nodes can make it very difficult to separate recycled ICs from a batch when the process variation outpaces aging degradation.

9.1.1 Simple RO-CDIR

A different approach was proposed in [35] based on ring oscillators (ROs) that avoided the data collection altogether and applied a "self–referencing" concept to the measurement of use time. Specifically, [35] embeds two ROs within the chip and compares them to detect prior IC usage. The first RO is called the *reference RO*

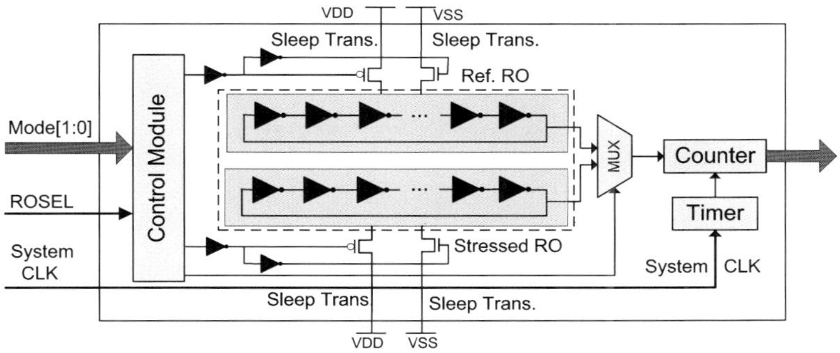

Fig. 9.2 Simple RO-CDIR sensor [35]

and is designed to age at a slow rate. The second RO is referred to as the *stressed RO*, and it is designed to age at a much faster rate than the reference RO. As the IC is used in the field, the stressed RO's rapid aging reduces its oscillation frequency while the reference RO's oscillation frequency remains largely static over the chip's lifetime. Thus, a large disparity between the two ROs' frequencies implies that the chip has been used. To overcome global and local process variations, the two ROs are placed physically very close together so that the process and environmental variations between them are negligible.

Figure 9.2 shows the structure of this simple RO-CDIR, which is composed of a control module, a reference RO, a stressed RO, a MUX, a timer, and a counter. The counter measures the cycle count of the two ROs during a time period controlled by the timer. The system clock is used in the timer to minimize the measurement period variations due to circuit aging. The MUX selects which RO is going to be measured and is controlled by the ROSEL signal. The inverters in the ROs can be replaced by any other types of gates (NAND, NOR, etc.) only if they can construct a RO. It will not change the effectiveness of the RO-CDIR significantly, according to prior analysis in [35]. In 90 nm technology, a 16-bit counter can operate at a frequency of up to 1 GHz, which means that an inverter-based RO must be composed of at least 21 stages [35].

9.1.2 *Limitations of Simple RO-CDIR*

Given the objective for designing RO-CDIR, the best RO-CDIR sensor (i.e., the one that detects recycled ICs most accurately) should possess minimal aging for the reference RO and maximum aging for the stressed RO. This cannot be achieved by the RO-CDIR proposed in [35] because, in the RO-CDIR design shown in Fig. 9.2,

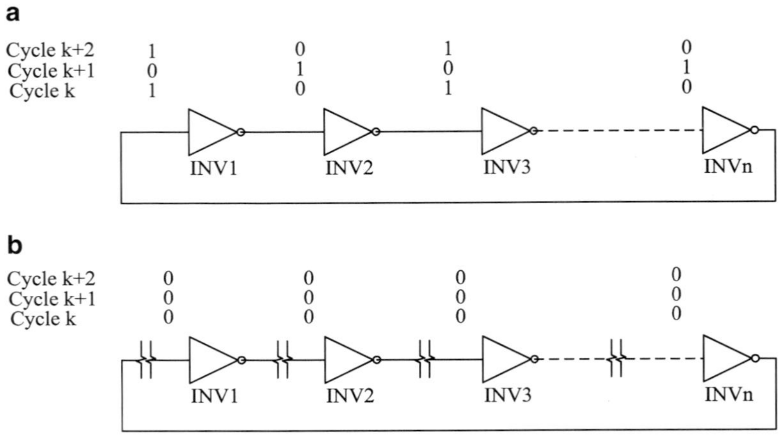

Fig. 9.3 NBTI stress on stressed ROs. (**a**) Stressed RO in RO-CDIR sensor [35] in stress mode. (**b**) Stressed RO in our proposed NBTI-Aware RO-CDIR sensor in stress mode

only half of the inverters in the stressed RO are negative-bias temperature instability (NBTI) stressed in one oscillation cycle as shown in Fig. 9.3a. This implies that half of the inverters age while the other half recover some of their aging. For example, at cycle time k, the even number inverters (e.g., inverters $2, 4, 6, \ldots$) are stressed as they receive zero at their inputs (zero causes the PMOS to age) while odd number inverters (e.g., inverters $1, 3, 5, \ldots$) recover their aging. At cycle time $k + 1$, the even number inverters recover and odd number inverters age. This process continues during normal operations and results in a slower aging for the stressed RO because the PMOS transistors partially recover every other cycle. Hot carrier injection (HCI) will not contribute as much to the total degradation of this sensor in the field since the sensors are kept in non-oscillatory mode. More details on the aging and recovery process can be found in [37, 38].

This problem is overcome in [36] where all the inverters are NBTI stressed during the entire operation of an IC where the RO-CDIR is deployed. Figure 9.3b shows the proposed solution where all the inverters are stressed during normal operations. This is achieved by breaking the connection of each inverter to its prior one and pulling down their inputs to ground. NBTI stress occurs when the gate of a PMOS transistor is pulled down to ground. Thus, all the inverters of the stressed RO are NBTI stressed during the entire time of the operation. As a result, the aging recovery cannot take place. However, if the chip is completely powered off, a partial recovery may occur. Nevertheless, the permanent degradation is proven to be much larger than the recovery [39].

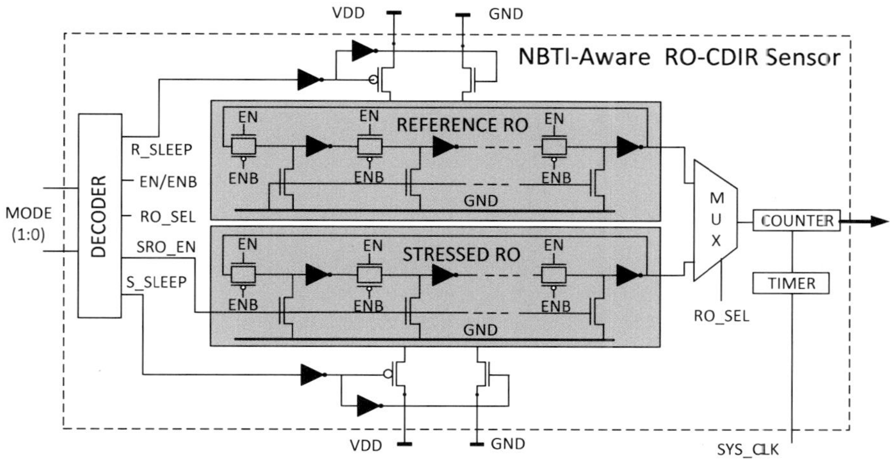

Fig. 9.4 The proposed NBTI-Aware RO-CDIR sensor

9.1.3 Design and Operation of NBTI-Aware RO-CDIR

Figure 9.4 shows the design of the NBTI-Aware RO-CDIR sensor [36]. The stressed RO is modified in such a way that all the inverters are stressed constantly during normal operation, as explained above. To achieve this, a pass transistor is introduced in between every pair of inverters, and the inputs of all the inverters are pulled down to ground using an NMOS network. To match all the internal parameters (node capacitance, resistance, etc.), the same pass transistor and NMOS are mimicked in the reference RO. This is to ensure that at time 0, when there is no aging, the difference between the two ROs is minimal and is mainly impacted by the manufacturing process variations present between the two ROs. A decoder is introduced to generate all the internal signals for a specific mode. When $EN = 0$, both ROs oscillate while the sleep transistors are ON. The signals *EN* and *SRO_EN* can never be "1" simultaneously as they would create a short circuit in the design. Similar to the design described in Fig. 9.2, the NBTI-aware RO-CDIR also has a MUX, a counter, and a timer to select the ROs and measure their frequencies during authentication. Also, sleep transistors are used to connect the ROs to the power supply in the RO-based sensor as before. PMOS sleep transistors control the connection between *VDD* and the inverters and NMOS sleep transistors control the connection between *VSS* and the inverters.

Table 9.1 highlights the four distinct modes of operation. In the manufacturing and burn-in tests, our objective is to protect both ROs from aging. In this mode, both ROs enter sleep mode by being cut off from the power and ground lines. *R_SLEEP* and *S_SLEEP* are assigned to "0" during this entire operation. In normal operation, the reference RO remains in the sleep mode while the stressed RO is in the stressed mode. All the inverters in the stressed RO are given a DC stress by pulling their

Table 9.1 Modes of operation

Mode	Signals *R_SLEEP*	*EN*	*RO_SEL*	*SRO_EN*	*S_SLEEP*	Description
00	0	X	X	X	0	Manufacturing and burn-in tests: both ROs are in sleep mode
01	0	0	X	1	1	Normal operation: reference RO in sleep mode and stressed RO in stressed mode (inverter input GND)
10	1	1	0	0	1	Authentication mode: measure frequency of reference RO
11	0	1	1	0	1	Authentication mode: measure frequency of stressed RO

inputs to ground. In authentication mode, the reference RO is activated to measure its frequency (*RO_SEL* to 0), which corresponds to the RO frequency of a new IC. Then, the stressed RO is activated (*SRO_EN* to 0 and *EN* to 1) and its frequency (*RO_SEL* to 1) is measured.

9.1.4 Overhead Analysis

The area overhead of both the RO-CDIRs is negligible for modern designs. The area overhead mostly comes from the size of the counter and timer. The area of the remaining parts is negligible. Thus both the original and NBTI-aware designs offer similar area overhead. We can also remove the timer and counter from the RO-CDIRs and measure the frequencies off-chip making the area overhead even smaller.

Table 9.2 shows the area overhead analysis of the RO-CDIRs. We define area overhead as the ratio of the size (area) of the RO-CDIR with the size (area) of the benchmark. Here, the IWLS 2005 benchmarks are arranged from low to high sizes to compute the area overhead. The timer and counter are excluded during the computation, as we assume the frequency measurement can be performed off-chip. As seen, the overhead is more than 1 % for small benchmarks (*i2c, spi, and b14*) for 51-stage NBTI-Aware RO-CDIR that could make it challenging to use them in small designs. The area overhead for the 51-stage RO-CDIR is less than the 51-stage NBTI-Aware RO-CDIR. The area overhead is comparably lower for the 21-stage RO-CDIRs. For large designs, however, it hardly impacts the overall area overhead.

The power consumption of the NBTI-Aware RO-CDIR is lower compared to the simple RO-CDIR, as there is no switching during the normal operation due to the fact that all inputs of the inverters in the stressed RO are pulled down to ground. However, both of them provide negligible power overhead when they are placed in modern industrial designs. As shown in Fig. 9.1, RO-CDIRs are suitable for large

Table 9.2 Area overhead analysis of RO-CDIRs

Benchmark	Size (# Gates)	Area overhead Simple 21-stage RO-CDIR (%)	NBTI-Aware 21-stage RO-CDIR (%)	Simple 51-stage RO-CDIR (%)	NBTI-Aware 51-stage RO-CDIR (%)
i2c	1, 124	2.89	4.73	5.52	9.98
spi	3, 277	1.01	1.65	1.92	3.48
b14	8, 679	0.38	0.62	0.73	1.31
b15	12, 562	0.26	0.43	0.50	0.91
DMA	19, 118	0.17	0.28	0.33	0.6
DSP	32, 436	0.10	0.17	0.19	0.35
ethernet	46, 771	0.07	0.115	0.135	0.244
vga_lcd	124, 031	0.03	0.044	0.051	0.092
leon2	780, 456	0.004	0.007	0.008	0.015

Table 9.3 Process variations

Process variations	Inter-die			Intra-die		
	Vth(%)	L(%)	Tox(%)	Vth(%)	L(%)	Tox(%)
PV0	5	5	2	5	5	1
PV1	8	8	3	7	7	2
PV2	20	20	6	10	10	4

digital ICs such as microprocessors, microcontrollers, digital signal processors, ASICs, programmable logic devices, and memories. Such sensors can also be used in smaller digital ICs if the area overhead is acceptable.

9.1.5 *Simulation of the NBTI-Aware RO-CDIR*

In order to verify the effectiveness of the NBTI-Aware RO-CDIR, the design is implemented and simulated using the 90 nm technology node [40]. HSPICE MOSRA from Synopsys is used to simulate and measure the impact of aging on this RO-CDIR. The nominal supply voltage is 1.2 V. In this simulation, we select 21-stage and 51-stage ROs to compare the results. To model the variation, Monte Carlo (MC) simulation is performed with 1,000 samples of the NBTI-Aware RO-CDIR in HSPICE. Here, we were mostly concerned with detecting ICs used in the field for a very short period of time, so we set, the total aging time at 15 days in the increment of 3 days. Larger usage times would be easily detected using this sensor.

Three different process variations are considered to investigate the impact of variation on the detection of recycled ICs. Table 9.3 shows the different process variations used in the simulation. Moving from PV0 to PV2, inter-die and intra-die variations both become larger. That is because, as feature size decreases and die size increases, the complex semiconductor manufacturing processes cause variations to

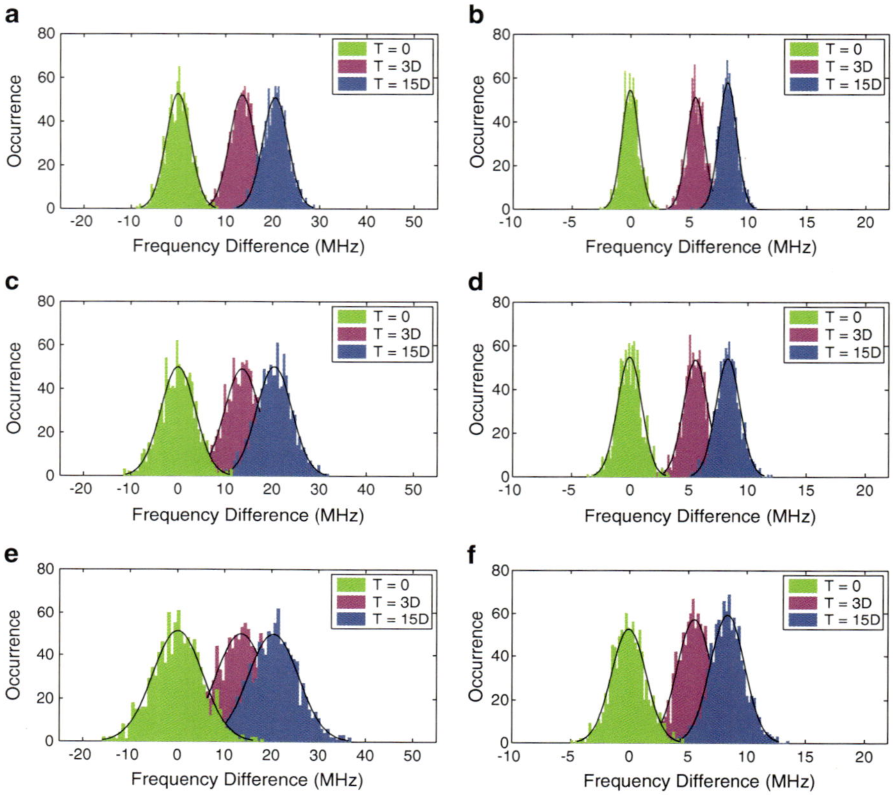

Fig. 9.5 The distribution of frequency differences between the reference RO and the stressed RO with different process variations, PV0, PV1, and PV2. (**a**) PV0: 21 Stage RO. (**b**) PV0: 51 Stage RO. (**c**) PV1: 21 Stage RO. (**d**) PV1: 51 Stage RO. (**e**) PV2: 21 Stage RO. (**f**) PV2: 51 Stage RO

the device parameters significantly. However, we acknowledge that the impact of process variation on ROs will be minimal as they are placed physically near to each other. PV0 represents the expected process variation between ROs while the other two are the worst-case scenarios.

Figure 9.5 shows the simulation results for the NBTI-Aware RO-CDIR sensor. The x-axis represents the frequency difference ($f_{diff} = f_{reference\_ro} - f_{stressed\_ro}$) between the reference RO and the stressed RO. The y-axis represents the frequency of occurrence (i.e., # of Monte Carlo samples). The legend in the figures denotes the aging time (for example, T = 3D denotes the RO-CDIR is aged for 3 days). The green distribution represents the f_{diff} distribution for the new ICs where the RO-CDIR has yet to be aged and is centered at 0 MHz. The pink and blue distributions represent 3 days and 15 days of aging respectively. It is clear that aging shifts the distributions to the right as the stressed RO has aged more and become slower resulting in the right shift of f_{diff} distribution. Also it is very important to observe the spreading of the distribution as the misprediction rate (will be introduced shortly) directly depends on it. The spreading becomes larger when the process variations become larger among the ICs. This is clearly shown in Fig. 9.5a,c,e and Fig. 9.5b,d,f.

We can clearly identify recycled ICs when the two distributions (T $=$ 0 and $T = 3, 15$ D) do not overlap with one another. The percentage of misprediction (new ICs detected as counterfeit and vice versa) can be estimated as the area of overlap between these two distributions. We apply Gaussian fit to find the mean and variance of the distributions and then calculate the overlapped area. We can certainly identify recycled ICs that have been aged more than 15 days in almost all cases. Based on this figure, we expect a higher misprediction rate (1) as the process variation increases and (2) when the 21-stage RO is used rather than the 51-stage RO. As process variation increases, the variance in f_{diff} grows, which results in a larger overlap between 0D and 3,15D distributions. Similarly, since the 21-stage RO distributions have a larger spread than the 51-stage RO, we should also expect higher misprediction rates. The best-case scenario occurs for the 51-stage RO with PV0 where we can detect recycled ICs in 3 days with negligible risk of misprediction. *This represents a substantial improvement over the prior work [35] which required at least 1 month of aging to identify recycled ICs*. As described in Sect. 9.1, 50 % of inverters in the stressed RO in simple RO-CDIR age in each oscillation cycle while the other half of inverters recover. This results in a slower aging of the stressed RO. In contrast, the inverters in the stressed RO in NBTI-Aware RO-CDIR age constantly (without recovering) during normal operation. Thus, we expect higher aging for the stressed RO, which allows our NBTI-aware RO-CDIR to detect recycled ICs used much less than 1 month (as little as 3 days).

9.1.6 Misprediction Rate Analysis

In order to find the effectiveness of RO-CDIR, we present the misprediction rate analysis. We define misprediction rate as recycled ICs identified as new ($\triangle_1$), and new ICs identified as recycled ($\triangle_2$). Here we will only present the results for NBTI-Aware RO-CDIR. Figure 9.6 shows the two distribution functions of the new and aged ICs having 21-stage ROs (aged for 3 days with PV2 process variation—worst case). The x-axis represents the frequency differences between the two ROs (f_{diff}) and the y-axis represents the corresponding distribution function. The overlap area represents the misprediction rate for identifying new or recycled ICs. The decision threshold should be the point (x_{th}) where both distributions intersect each other. The green area represents the probability of identifying new ICs as recycled whereas the red area denotes the probability of identifying recycled ICs as new. These areas ($\triangle_1$ and $\triangle_2$) are represented by:

$$\triangle_1 = \int_{x_{th}}^{\infty} f_{0D}(x)\,\mathrm{d}x, \; \textit{and} \; \triangle_2 = \int_{-\infty}^{x_{th}} f_{nD}(x)\,\mathrm{d}x$$

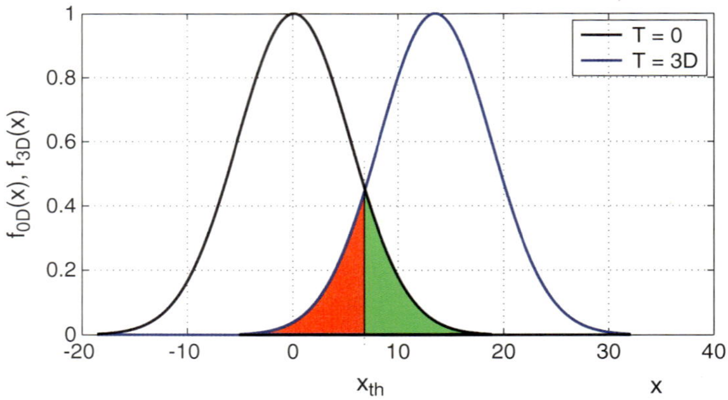

Fig. 9.6 Probability density function of frequency differences between 21-stage reference and stressed ROs

where, f_{0D}, and f_{nD} corresponds to the distribution of frequency differences for new ICs and ICs with n days of aging, respectively.

Table 9.4 shows the misprediction rate i.e., recycled ICs identified as new (Δ_1) and new ICs identified as recycled (Δ_2) for 21-stage and 51-stage NBTI-Aware RO-CDIR, with the process variations mentioned in Table 9.3. The rate is higher in PV2, as stressed and reference ROs' frequencies differ significantly between the two samples due to their higher process variations. This results in a larger area of overlap between the two distributions. However, we obtain significantly lower Δ for 51-stage RO. Δ_1 is 2.79 % and 0.21 % when the NBTI-Aware RO-CDIR was aged 3 days and 15 days, respectively. For PV1, it is 0.32 % and 0 % for same use times. We were able to predict whether they were recycled or new even though they were aged only 3 days with a small percentage of error. As we described earlier, with these two ROs placed very close to each other, the variation should be well below PV1 and we should predict all the samples. Under different cases, we also observed a similar misprediction rate (Δ_2) of identifying new ICs as recycled. In both these misprediction (Δ_1 and Δ_2) cases, the 51-stage RO outperforms the 21-stage RO.

In the simulation, we have only considered process variation. We did not include any results for temperature and power supply variation. As the two ROs are placed very close in the circuit layout and the temperature variation is a global phenomenon, the temperature variation between the two is practically negligible ($\Delta T = 0$). At higher temperatures, we would also expect more rapid aging in the stressed RO, which should only improve our results. A similar argument can be made for power supply variation.

Table 9.4 Misprediction rate for NBTI-Aware RO-CDIR

	Recycled ICs identified as new (Δ_1)						New ICs identified as recycled (Δ_2)					
	3 days			15 days			3 days			15 days		
	PV0(%)	PV1(%)	PV2(%)	PV0(%)	PV1(%)	PV2(%)	PV0(%)	PV1(%)	PV2(%)	PV0(%)	PV1(%)	PV2(%)
21-stage RO	0.6	3.53	10.19	0	0.31	2.84	0.45	3.16	10.54	0	0.25	2.87
51-stage RO	0	0.32	2.79	0	0	0.21	0	0.3	2.85	0	0	0.18

Table 9.5 Workload analysis

	Workload				
	100 %	75 %	50 %	10 %	1 %
51-Stage RO-CDIR	3 days	4 days	6 days	30 days	300 days

9.1.7 Workload Analysis

It is also important to analyze different workloads that impact the detection of recycled ICs. We define workload as the percentage of time per day that the IC is in use. The workload/usage depends on the type of application being run. For example, the ICs used in—(1) mobile phones may remain on during the entire day (workload may be 100 %), or, (2) televisions or laptops may be ON for a fraction of day (workload may be well below 100 %). *We have considered 100 % workload for all the simulations unless specified otherwise.* Table 9.5 shows the minimum usage time of ICs under various workloads required for proper identification. Note we have shown the results only for the 51-stage NBTI-Aware RO-CDIR, as it provides minimum misprediction. The results show that the length of time required to detect the recycled IC increases as the workload decreases. For example, a workload of 10 % and 1 % requires the IC be used for 30 days and 300 days respectively. With reduced workload, we can only identify ICs as recycled if the system is used over a longer period of time because when the system is off (i.e., not in use), time passes, but the stressed RO does not age at all. Note that the impact of low-workload environment would be similar for all prior approaches based on aging [9, 10, 35]. Hence, the NBTI-aware RO-CDIR will outperform all other aging-based methods at any workload.

9.1.8 Attack Analysis

As we all know, counterfeiting is an evolving problem. The counterfeiters are continuously improving their techniques through experience. We believe that this trend will continue and the counterfeiters will continue to evolve and adapt their techniques to new detection and protection methods. Thus, it is of the utmost importance to analyze all of the possible attacks on these RO-CDIRs and their vulnerabilities in order to examine their robustness. There may be two types of attacks possible on RO-CDIRs, and they are as follows:

- *Removal/Tampering*: The first attack on RO-CDIRs could be removal/tampering attacks. However, it is fairly impossible for the counterfeiter to replace the stressed RO with a new one or to tamper with the stressed/reference RO in order to match their frequency. If we assume that a removal or tampering attack is possible, then the counterfeiter must remove the old package and then again repackage and remark it according to its original specifications. This removal and then repackaging may not be cost effective to the counterfeiters. Hence, it is unlikely to be used in practice.

- *Age Reference RO*: In this attack scenario, the counterfeiter may try to intentionally age the Reference RO to mask the difference between the ROs. The counterfeiter might attempt to force the RO-CDIR to work in authentication mode (MODE 10, in Table 9.1) for a period of time under accelerated stress conditions. With the accelerated aging at the same time, the frequency difference between the Stressed RO and the Reference RO would shrink since both of them could asymptotically approach maximum degradation.
  As we all know, burn-in is a very expensive process and the counterfeiter must have an expensive setup for that. The primary incentive for counterfeiting is cheap recycling, not adding extra cost to the components. There might not be any motivation left for the counterfeiters when they are forced to add burn-in to their recycling process. As a result, this attack might not be feasible as there is no cost incentive.

9.2 Antifuse-Based CDIR Structures

The NBTI-Aware RO-CDIR sensor described above approximates the aging time in terms of the frequency difference between the reference and stressed ROs. Even though the ROs are placed physically next to each other, there might be some intra-die process variations that limit the resolution of the RO-CDIR, especially if the device is kept OFF more than it is ON. Another limitation of the RO-CDIR sensors is their dependence on aging mechanisms in lower technology nodes. In other words, such sensors may not be well suited for older technology nodes because of limited aging, making them less attractive for use in military and space applications which usually use older technology nodes because of reliability concerns. The simulation results (see Sect. 9.1.5) for the NBTI-Aware RO-CDIR show that the detection of aged ICs under process corners is possible if they are used continuously for at least 3 days. To address the accuracy issue, an antifuse-based CDIR (AF-CDIR) structure was proposed in [41] and was primarily targeted at large digital ICs. In this section, we will first describe antifuse memory and then present two different AF-CDIR structures, namely, CAF-based CDIR and SAF-based CDIR.

9.2.1 Antifuse Memory

An antifuse (AF) is an electronic device that changes state from high resistance and non-conductive to low resistance and conductive in response to electrical stress. With sufficiently high voltage/current, large power dissipation in a small area will melt a thin insulating dielectric between polysilicon and diffusion electrodes and

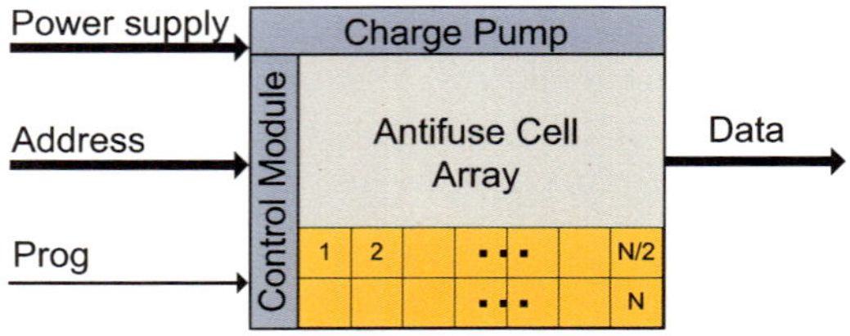

Fig. 9.7 Typical interface of antifuse memory [41]

form a thin, permanent, and resistive silicon link. The programming performed after manufacturing is irreversible and permanent in antifuse cells, which will be used in AF-CDIRs to store the usage time of ICs. The reasons [42] for using an antifuse block in the AF-based sensor are (1) it consumes less power to program or read compared with other types of OTP structures, such as electrical fuse or CMOS floating gate; (2) the area of an antifuse is much smaller than that of an efuse; and (3) it does not require additional masks or extra handing steps during fabrication.

Most antifuse memories are programmed in a programming environment with relatively high voltage or current. Therefore, integrated charge pumps or voltage multipliers are used to provide sufficiently high voltage or current [43, 44] in embedded antifuse OTP memories. With those charge pumps or voltage multipliers, no additional power supply is required during programming. The typical interface for the embedded antifuse memory is shown in Fig. 9.7, including *Power supply, Address, Prog*, and *Data* signals. We use existing antifuse blocks with the interface shown in Fig. 9.7 in the AF-based sensors.

9.2.2 Clock AF-Based (CAF-Based) CDIR

Figure 9.8 shows the structure of the CAF-based CDIR, which is composed of two counters, a data read module, an adder, and an antifuse OTP memory block. *Counter1* is used to divide the high frequency system clock to a lower frequency signal. *Counter2* is used to measure the cycle count of the lower frequency signal. The size of the two counters can be adjusted depending on the measurement scale (T_s: defined as the time unit reported by the sensor) and the total measurement time (T_{total}). Here the size of *Counter1*, and *Counter2* depend on T_s, and T_{total}/T_s respectively. Let us start with an example for calculating the size of *Counter1* and *Counter2*. Assume that T_s is 1 h and T_{total} is 1 year. The system clock frequency ($f = \frac{1}{T}$) is 50 MHz. Now the maximum count for *Counter1* and *Counter2* are

$$Count1_{max} = \frac{1\,h}{clock\ cycle\ time} = \frac{3600}{T} = 3600 * f = 3600 * 50 * 10^6\ (> 2^{37} \& < 2^{38})$$

$$and$$

$$Count2_{max} = \frac{365 * 24\,h}{1\,h} = 8760\ (> 2^{13} \& < 2^{14}) \tag{9.1}$$

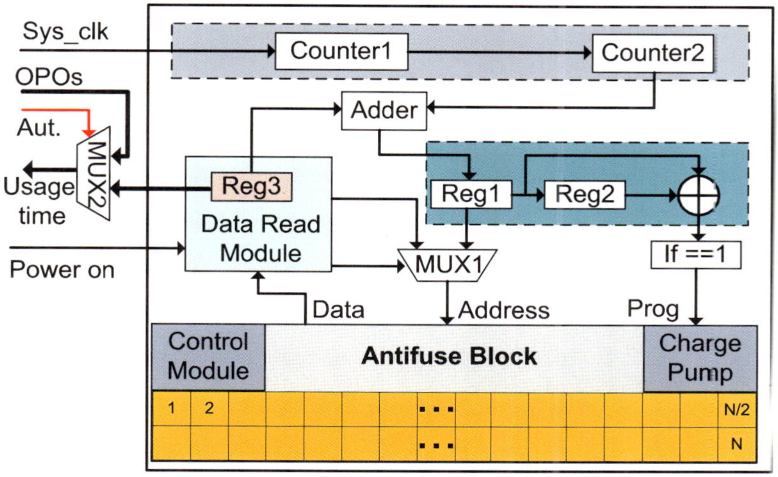

Fig. 9.8 The structure of the CAF-based sensor

From the above Eq. 9.1, it is obvious that the size of *Counter1* and *Counter2* will be 38 bit and 14 bit respectively. It is also clear that the size of *Counter1* depends on the system clock (Sys_clk). If the design uses multiple clocks, then we will select the slowest clock, as it minimizes the *Counter1* area which is directly related to the count to reach T_s.

An embedded antifuse OTP block is used to retain the data permanently. We use antifuse OTP as it provides a lower cell area and an improved tamper resistance over other technologies. The design keeps track of *Counter2* output to retain possession of usage time. *Prog* is assigned to be $1'b1$ if the value of *Counter2* increases by "1". By connecting the output of *Counter2* to *Address* in the antifuse block directly, the related antifuse cell will be programmed as "1". Therefore, the largest address of the cell whose content is "1" will be the usage time of the IC based on the measurement scale setup by *Counter1*.

However, program and read operations share the same $Address$ signals in antifuse block. Therefore, a multiplexer (*MUX1* in Fig. 9.8), controlled by *data read module*, is used to select the address of the antifuse cell to be read or programmed. Every time the power supply is on, the antifuse block will work in read mode for a short period of time. During this time, the read address generated by *data read module* will go through *MUX1*, and all the antifuse cells will be traversed based on the traversing binary tree principle. Figure 9.9 shows the algorithm for data read in a N-bit antifuse block. From Fig. 9.9, we can see that there are $log(N/2)$ loops in the algorithm. The address is increased or decreased by $2^{i-1}(i = 0, \dots log(N/2))$ for the ith loop based on the value in the address. If the value stored in the address is "1" ($[address] == 1$) and the value stored in the next address is "0", the address

Fig. 9.9 Algorithm for "data read" in CAF-based and SAF-based sensors

```
Algorithm for Data Read
initial address = (N/2);
for (i = log(N/2), i > 0, i − −) {
    if ([address] == 1)
        address = address + 1;
        if ([address] == 0)
            address = address − 1, $stop;
        else
            address = address − 1;
        address = address + 2^(i−1);
    else
        address = address − 2^(i+1);
}
```

will represent the usage time before power-on based on T_s. The read operation will last less than $log(N/2) + 1$ system clock cycles, depending on the value stored in the antifuse block; this time will be recorded by *Counter1*, as well.

Once we get the previous usage time, it will be stored in the register *Reg3* and sent to the $adder$. The reason for using an adder here is that the counters start from "0" every time the power is turned on and the previous usage time must be considered when we calculate the total usage time. In addition, *Reg1* is used to sample the data in *adder*, *Reg2* delays the data in *Reg1* with one system clock, and *XOR* gates are used to compare the data in *Reg1* and *Reg2*. If they are different (denoting the usage time increased), the antifuse OTP block will work in program mode and the data in *Reg1* will go through *MUX1* to the $Address$ in the antifuse block. Therefore, combined with the value in *Counter2* (the usage time after power-on), the new total usage time will be stored in the antifuse OTP block by programming a new antifuse cell with a larger address. From the above discussion, we can see that the antifuse OPT block is programmed internally. By designing our sensor in this way, we can reduce the probability of altering or tampering attacks on the CAF-based CDIR.

In order to eliminate the need for additional pins for authentication purposes on the chip, the CAF-based CDIR uses a multiplexer (*MUX2*) and an authentication (*Aut.*) pin to send the usage time to the output pins of ICs. This way, no extra output pins will be added to the original design. Thus, this AF-CDIR requires only one extra pin. When the IC works in normal functional mode, original primary outputs (*OPOs*) will go through *MUX2*. If the IC is in authentication mode by enabling the authentication signal, the *data read module* will set the antifuse IP in read mode, and the usage time will go through *MUX2*. In addition, when the IC works in manufacturing test mode, the functionality of the CAF-based CDIR will be disabled and structural fault test patterns will be applied to this structure.

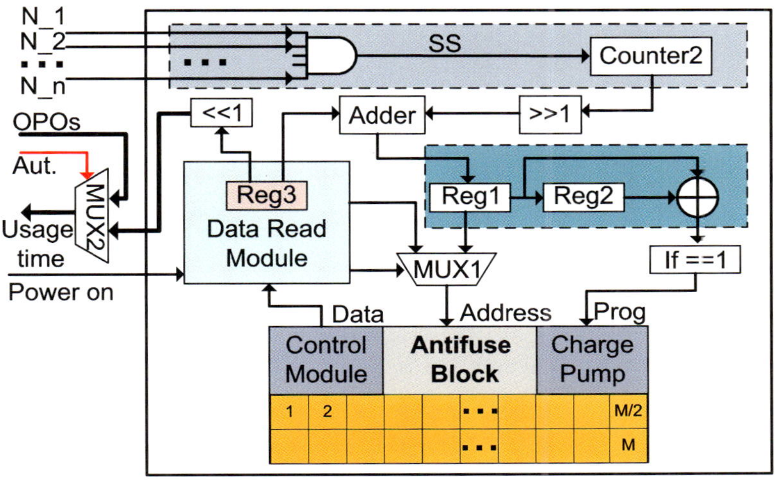

Fig. 9.10 The structure of the SAF-CDIR sensor [41]

9.2.3 Signal AF-Based (SAF-Based) CDIR

With two counters, the area overhead of the CAF-based sensor could still be considered large for smaller designs. In order to reduce the area overhead, the SAF-based CDIR is proposed which is based on the switching activity (SW) of a few internal nets of an IC. The SAF-based CDIR's structure is similar to that of CAF-based CDIR. However, the difference is that the CAF-based CDIR counts the cycles of a system clock to record the usage time of an IC while the SAF- based CDIR counts the switching activity (positive edge) of a certain number of nets in an IC.

Figure 9.10 shows the structure of the AF-CDIR, which is composed of a counter, a data read module, an adder, and an antifuse OTP memory block. *Counter2* is used to count the switching activity (positive edges) of a certain number of nets in the design. Since a field programmable read-only memory (FPROM) could be tampered with or altered by attackers, an embedded antifuse OTP block is used to store the usage time (total number of positive edges). Program and read operations share the same $Address$ in an antifuse block. Therefore, a MUX ($MUX1$), controlled by data read module, is used to select the address (antifuse cell) to be read or programmed.

Every time the power supply is ON, the antifuse block will work in read mode for a short period of time. During this time, the read address generated by the data read module will go through $MUX1$ and all the antifuse cells will be traversed based on the traversing binary tree principle, described in Fig. 9.9 algorithm.

A 1-bit right shifter is used to divide the value in *Counter2* by 2 and then the largest address of antifuse cells with "1" will represent *[SW/2]* in order to reduce the area overhead. A 1-bit left shifter is used to calculate the switching activity by

Table 9.6 Area overhead for CAF-based, and SAF-based sensors on CSAFTEST

Measurement		Area overhead			Area of
Scale (T_s)	Total time (T_{total})	CAF-based(%)	SAF-based(%)	Reduction(%)	CSAFTEST(%)
1 min	1 month	7.37	3.72	49.5	500 K gates
1 h	1 year	1.57	0.82	47.8	and
1 day	1 year	0.18	0.12	33.3	12 KB
1 day	4 years	0.37	0.21	43.2	memory

*[SW/2] * 2*. The recorded *SW* will represent the ICs' usage time. Therefore, the number of antifuse cells in the SAF-based sensor will be reduced compared with the CAF-based sensor. However, the accuracy of the SAF-based sensor is lower than CAF-based sensor because (1) it is based on the switching activity of a certain number of nets in the netlist, while the CAF-based sensor counts the cycle count of the system clock, and (2) the SAF-based sensor loses part of the usage time information due to the shifters.

9.2.4 Area Overhead Analysis

In order to verify the effectiveness of AF-CDIRs, we analyzed the area overhead on the implementation of a design (we name it as CSAFTEST) with about 500 K gates and 12 KB in-system programmable memory. Table 9.6 shows the area overhead caused by CAF-CDIR, and SAF-CDIR, with different measurement scales and total measurement time. From the table, we can see that the area overhead caused by AF-CDIRs change with T_s and T_{total} since their structures change with measurement resolutions. For CAF-CDIR, the size of *Counter1* depends on T_s while the size of *Counter2* and the size of the antifuse memory block both depend on T_{total}/T_s. For SAF-CDIR, the area overhead is much smaller than that of CAF-CDIR due to the omission of *Counter1*. The reduction of overhead is calculated by the following formula,

$$Reduction = \frac{Overhead(CAF - based) - Overhead(SAF - based)}{Overhead(CAF - based)} X 100\,\%$$

This reduction is shown in the fifth column in Table 9.6. For example, with T_s=1 h and T_{total}=1 year (8,760 h), CAF-CDIR was designed with 20-bit *Counter1*, 14-bit *Counter2*, and 8,760-bit antifuse memory block. The area overhead of this CAF-CDIR is 1.57 % while the area overhead caused by SAF-CDIR is 0.82 % and the reduction is 47.8 %. However, if T_s = 1 min & T_{total} = 1 month and T_s = 1 day & T_{total} = 1 year, the area overhead of CAF-CDIR are 7.37 % and 0.18 %, respectively.

From the above analysis, we can see that the area overhead caused by AF-CDIRs depend on the application and specification of ICs. For example, if an IC is used in a system that requires a small T_s and a large T_{total}, the area overhead would be large. Otherwise, the overhead would be small (less than 1 %). is power-on time and the intervals between power-on are not calculated. Therefore the usage time stored in the sensor (T_{total}) is usually shorter than the time with power-off intervals. With a smaller T_{total}, the size of the antifuse memory block in our AF-based sensors will be smaller and accordingly the area overhead will be smaller.

9.2.5 *Attack Analysis*

AF-CDIRs are inherently resistant to attacks as there is no control access to the CDIRs. The counterfeiter can only access the *Aut.* pin to read out the usage time. We will still analyze all the possible attacks that can be performed on it. All the possible scenarios are as follows:

- *Tampering*: For AF-based sensors, attackers could try to mask the usage time of the ICs by disabling the CDIR. However, the AF-based sensor will automatically run whenever power is on, and the usage time will be stored in the antifuse memory directly. Therefore, it is impossible for attackers to disable the CDIR without removing the package and breaking the chip.
- *Erasure of antifuse memory*: The second attack could be the erasure and alteration of antifuse cells; this is not possible because the memory used in our sensors is an antifuse OTP block. The most important advantage of the antifuse OTP technique is its ability to resist all existing reverse engineering methods because the oxide breakdown in antifuse cells occurs in a random location within a bounded enclosure and is extremely small [42]. Therefore, the state of a bit cell stays well hidden in the silicon atoms, which makes it extremely difficult for attackers to tamper with the memory.
- *Modification of counter content*: The third attack could be the modification of counters or signals connection in the sensor. However, with limited resources and without access to the original design, attackers cannot modify the nets connection.

9.3 Fuse-Based CDIR

The RO-CDIR and AF-CDIR structures describe above, are most suitable for large digital ICs due to the area required to implement them. However, the majority of components on the market today are smaller analog, digital, and mixed-signal types. In this section, we are presenting an alternative, low-cost structure that is based on semiconductor fuses [45, 46] and can be implemented into almost any design,

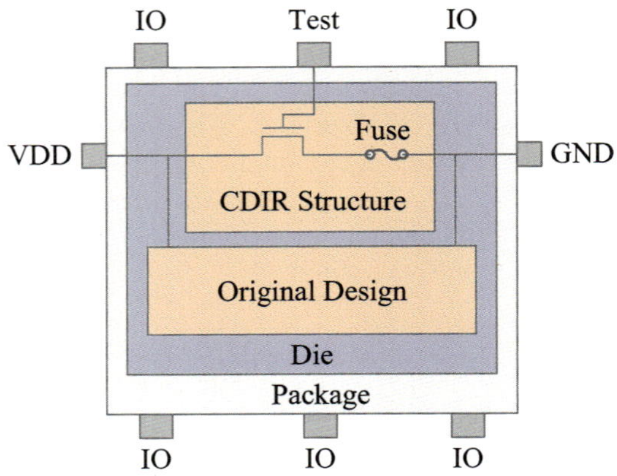

Fig. 9.11 F-CDIR: version I

with the exception of discrete components, such as diodes, transistors, and passive components. This structure can be fabricated along with the original design, and it does not require the modification or addition of any steps to the manufacturing process.

Figure 9.11 presents the design for the fuse-based CDIR structure. The structure consists of a switch and a fuse. It is a three-terminal structure, having two terminals that are connected to VDD and GND pins. The third terminal, the control terminal, is regulated by $Test$ pin on the IC. In this design, the MOSFET acts as the switch. The design overhead is only one transistor and a fuse. The design works as follows: During the manufacturing and burn-in test modes $Test$ pin will always be "0" which will provide no current flow through this structure. When the component is placed in the printed circuit board (PCB) for normal operation, $Test$ pin will be connected to VDD. The MOS will be ON and a current will flow through the fuse, which will result in an open circuit inside the structure. The device will then operate normally.

The detection of counterfeit (used in the field) components will be the measurement of resistance between VDD and GND pins while setting $Test$ pin to VDD. The measured resistance between VDD and GND should be negligible for new component. If the component has been used in the field, the measured resistance will be high (infinite). Here we are assuming the users of the component are trusted and they design the PCB with $Test$ treated as VDD. For the added security, the $Test$ pin can be named as VDD.

Figure 9.12 shows the implementation of this structure in differential designs. The structure is placed in between the differential output, $O+$ and $O-$, pins. The control pin is connected to the $Test$ pin. For the proper burning of the fuse, the differential design must provide the necessary current to the fuse. During the manufacturing and burn-in tests mode, $Test$ pin will be assigned to "0" which makes the MOS off and the fuse remains intact. When the device operates in field for the first time, the fuse will be burnt because of a current flowing through it.

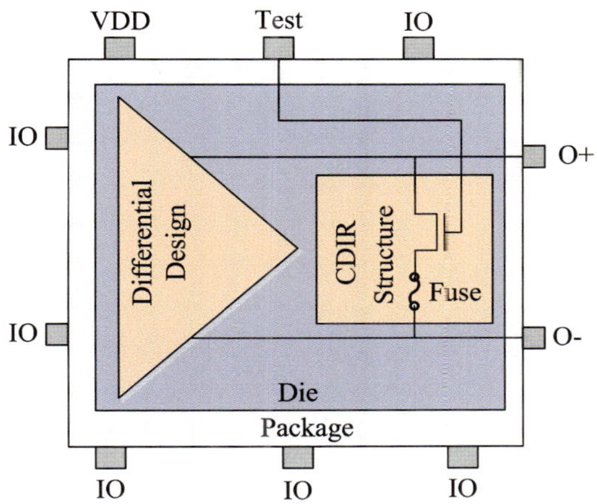

Fig. 9.12 F-CDIR version I implemented in differential designs

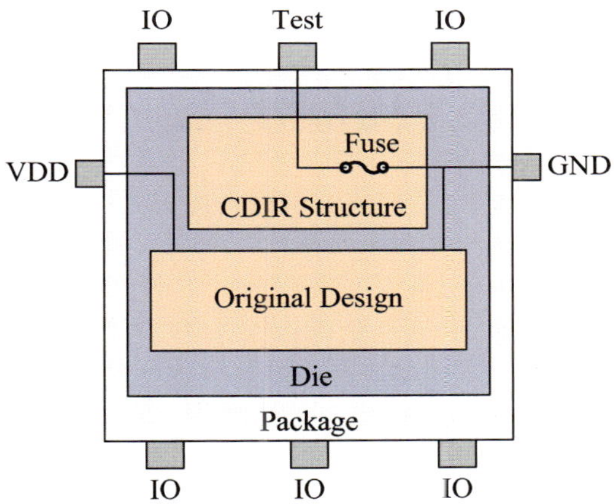

Fig. 9.13 F-CDIR: version II

The design will then operate according to its normal specifications. The measured resistance between $O+$ and $O-$ should be negligible for new components, and it will be high (infinite) for counterfeit components.

Figure 9.13 presents a different version of the CDIR structure. The design consists of only one semiconductor fuse. The terminals of the sensor are connected to $Test$ and GND pins. The fuse is isolated from the rest of the design. During the manufacturing and burn-in tests mode, the $Test$ pin will always be "0". The fuse will be intact during these modes, as there is no current flowing through it. In normal operation, this pin will be assigned to VDD. When the chip operates in the field for the first time, the current will flow through the sensor and the fuse will be burnt. The detection of used components will be based on measuring the resistance value between the $Test$ and GND pins. A simple multimeter can authenticate the components. A component will be treated as counterfeit if this measured resistance value is high (infinite) and new if this value is low.

Table 9.7 Area overhead of F-CDIR

Benchmark	Components	Area overhead	
		F-CDIR I(%)	F-CDIR II(%)
Operational amplifier #1	11	18.18	9.09
Continuous-time state-variable filter	42	4.76	2.38
Operational amplifier #2	10	20.00	10.00
Leapfrog filter	77	2.60	1.30
Digital-to-analog converter	44	4.54	2.27

Note that the successful implementation of F-CDIR relies on the trusted system integrator as the burning of the fuse will only be asserted if VDD gets applied to the $Test$ pin. If the system integrator does not apply VDD to the $Test$ pin in their systems, the fuse will then be intact. In that case, we cannot identify recycled ICs by simply measuring the resistance.

9.3.1 Area Overhead Analysis

Table 9.7 shows the approximate area overhead of the F-CDIRs. We have selected ITC'97 benchmark [47] for analog and mixed signal circuits. We have calculated the approximate area overhead by the ratio of components used in the F-CDIR with benchmark circuits. For small analog circuits, the overhead is about 20 % for the F-CDIR Version I, whereas it is considerably lower for the F-CDIR Version II. The F-CDIR II consists of only one component (fuse) compared to two components (a fuse and a transistor) for the F-CDIR I. As mentioned earlier, both F-CDIR structures also require one extra test pin. This might prohibit their use in cases where the number of IO pins are limited as they are in smaller ICs. As for digital circuits (such as the benchmarks used in Table 9.2), F-CDIR structures require virtually negligible area overhead and the one extra pin may not be an issue either.

9.3.2 Attack Analysis

The design for the F-CDIRs is the simplest among the all three CDIRs, as it consists of only one fuse (F-CDIR version II) or one fuse and one transistor (F-CDIR version I). However, this design is also resistant to tampering like the AF-CDIR design. The possible attacks are as follows:

- *Trust on System Integrator*: For the proper operation of the F-CDIR, burning the fuse is necessary, and this can only be done when VDD gets applied to the $Test$ pin. Thus the successful implementation of the F-CDIR relies on the trusted system integrator.

- *Tampering*: The state of the fuse could be modified. However, a separate metal deposition is necessary to make the fuse. This would require the decapsulation of the package and then metal deposition. This is indeed a very costly process. Thus there should not be any cost incentive for the counterfeiters to perform this process for every IC. The counterfeiters would not get any benefit, as these structures would be placed in very low-cost analog and mixed-signal ICs.

9.4 Summary

In this chapter, we have presented a set of DFAC structures, namely RO-CDIRs, AF-CDIRs, and F-CDIRs, to detect recycled and remarked ICs of different types and sizes. The NBTI-Aware RO-CDIR structure can be implemented in any digital IC with new technology nodes as it takes the advantage of higher aging in newer technology nodes. It can be placed even in smaller digital ICs with few thousand gates, due to the low area overhead. The simple RO-CDIR requires three test pins whereas the NBTI-Aware RO-CDIR needs two additional pins while also achieving better performance. The AF-CDIR can only be placed in large digital ICs. These ICs can be manufactured with new or older technology nodes as these AF-CDIRs function based on counting of system clock, or switching of internal nets. AF-CDIRs require only one additional pin. F-CDIRs can be implemented in any components (small, or large, and analog, or digital) and any technology node. These CDIRs can authenticate ICs very effectively and require a very low cost multimeter. F-CDIRs require only one test pin for IC authentication. Finally, all these CDIRs are resistant to all types of known attacks. Together, these structures provide excellent coverage for the full range of recycled ICs.

References

1. SAE, Counterfeit electronic parts; avoidance, detection, mitigation, and disposition, 2009. http://standards.sae.org/as5553/
2. IDEA, Acceptability of electronic components distributed in the open market. http://www.idofea.org/products/118-idea-std-1010b
3. CTI, Certification for coutnerfeit components avoidance program, September 2011
4. U. Guin, D. DiMase, M. Tehranipoor, Counterfeit integrated circuits: detection, avoidance, and the challenges ahead. J. Electron. Test. **30**(1), 9–23 (2014)
5. U. Guin, K. Huang, D. DiMase, J. Carulli, M. Tehranipoor, Y. Makris, Counterfeit integrated circuits: a rising threat in the global semiconductor supply chain. Proceedings of the IEEE **102** (8), 1207–1228 (2014)
6. U. Guin, D. DiMase, M. Tehranipoor, A comprehensive framework for counterfeit defect coverage analysis and detection assessment. J. Electron. Test. **30**(1), 25–40 (2014)
7. U. Guin, M. Tehranipoor, On selection of counterfeit IC detection methods, in *IEEE North Atlantic Test Workshop (NATW)*, May 2013

8. U. Guin, M. Tehranipoor, D. DiMase, M. Megrdichian, Counterfeit IC detection and challenges ahead, in *ACM/SIGDA E-NEWSLETTER*, vol. 43(3), March 2013
9. X. Zhang, K. Xiao, M. Tehranipoor, Path-delay fingerprinting for identification of recovered ics, in *Proceedings International Symposium on Fault and Defect Tolerance in VLSI Systems*, October 2012
10. K. Huang, J. Carulli, Y. Makris, Parametric counterfeit IC detection via Support Vector Machines, in *Proceedings International Symposium on Fault and Defect Tolerance in VLSI Systems*, 2012, pp. 7–12
11. IHS iSuppli, Top 5 most counterfeited parts represent a $169 billion potential Challenge for global semiconductor market, 2011
12. K. Arndt, C. Narayan, A. Brintzinger, W. Guthrie, D. Lachtrupp, J. Mauger, D. Glimmer, S. Lawn, B. Dinkel, A. Mitwalsky, Reliability of laser activated metal fuses in drams," in *Proceedings of IEEE on Electronics Manufacturing Technology Symposium*, 1999, pp. 389–394
13. N. Robson, J. Safran, C. Kothandaraman, A. Cestero, X. Chen, R. Rajeevakumar, A. Leslie, D. Moy, T. Kirihata, S. Iyer, Electrically programmable fuse (efuse): from memory redundancy to autonomic chips, in *IEEE Custom Integrated Circuits Conference CICC*, 2007 pp. 799–804
14. R. Pappu, Physical one-way functions. Ph.D. Dissertation, Massachusetts Institute of Technology, 2001
15. G. Suh, S. Devadas, Physical unclonable functions for device authentication and secret key generation, in *Proceedings of ACM/IEEE on Design Automation Conference*, June 2007, pp. 9–14
16. K. Kursawe, A.-R. Sadeghi, D. Schellekens, B. Skoric, P. Tuyls, Reconfigurable physical unclonable functions - enabling technology for tamper-resistant storage, in *Proceedings of IEEE International Workshop on Hardware-Oriented Security and Trust*, July 2009, pp. 22–29
17. S. Kumar, J. Guajardo, R. Maes, G.-J. Schrijen, P. Tuyls, Extended abstract: The butterfly puf protecting ip on every fpga, in *Proceedings of IEEE International Workshop on Hardware-Oriented Security and Trust*, June 2008, pp. 67–70
18. L. Bolotnyy, G. Robins, Physically unclonable function-based security and privacy in rfid systems, in *Proceedings of IEEE International Conference on Pervasive Computing and Communications*, March 2007, pp. 211–220
19. X. Wang, M. Tehranipoor, Novel physical unclonable function with process and environmental variations, in *Proceedings on Design, Automation Test in Europe Conference Exhibition (DATE)*, March 2010, pp. 1065–1070
20. Y.M. Alkabani, F. Koushanfar, Active hardware metering for intellectual property protection and security, in *Proceedings of 16th USENIX Security Symposium on USENIX Security Symposium*, 2007, pp. 20:1–20:16
21. K. Lofstrom, W. Daasch, D. Taylor, Ic identification circuit using device mismatch, in *Proceedings of IEEE International Solid-State Circuits Conference*, 2000, pp. 372–373
22. J. Lee, D. Lim, B. Gassend, G. Suh, M. van Dijk, S. Devadas, A technique to build a secret key in integrated circuits for identification and authentication applications, in *Proceedings of Digest of Technical Papers on VLSI Circuits*, June 2004, pp. 176– 179
23. Y. Su, J. Holleman, B. Otis, A 1.6pj/bit 96circuit using process variations, in *Proceedings of IEEE International on Solid-State Circuits Conference*, February 2007, pp. 406–611
24. F. Koushanfar, G. Qu, M. Potkonjak, Intellectual property metering, in *Information Hiding* (Springer, Berlin, 2001), pp. 81–95
25. R. Chakraborty, S. Bhunia, Hardware protection and authentication through netlist level obfuscation, in *Proceedings of IEEE/ACM International Conference on Computer-Aided Design*, November 2008, pp. 674–677
26. Y. Alkabani, F. Koushanfar, M. Potkonjak, Remote activation of ICs for piracy prevention and digital right management, in *Proceedings of IEEE/ACM International Conference on Computer-Aided Design*, 2007, pp. 674–677
27. J. Roy, F. Koushanfar, I. Markov, EPIC: Ending piracy of integrated circuits, in *Proceedings on Design, Automation and Test in Europe*, March 2008, pp. 1069–1074

28. J. Huang, J. Lach, IC activation and user authentication for security-sensitive systems, in *Proceedings of IEEE International Workshop on Hardware-Oriented Security and Trust*, June 2008, pp. 76–80
29. A. Baumgarten, A. Tyagi, J. Zambreno, Preventing IC piracy using reconfigurable logic barriers. IEEE Des. Test Comput. **27**(1), 66–75 (2010)
30. R. Chakraborty, S. Bhunia, HARPOON: an obfuscation-based SoC design methodology for hardware protection. IEEE Trans. Comput. Aided Des. Integr. Circuits Syst. **28**(10), 1493–1502 (2009)
31. G. Contreras, T. Rahman, M. Tehranipoor, Secure split-test for preventing IC piracy by untrusted foundry and assembly, in *Proceedings International Symposium on Fault and Defect Tolerance in VLSI Systems*, 2013
32. M. Miller, J. Meraglia, J. Hayward, Traceability in the age of globalization: a proposal for a marking protocol to assure authenticity of electronic parts, in *SAE Aerospace Electronics and Avionics Systems Conference*, October 2012
33. Semiconductor Industry Association (SIA), Public comments - DNA authentication marking on items in FSC5962, November 2012
34. C. Kuemin, L. Nowack, L. Bozano, N. D. Spencer, H. Wolf, Oriented assembly of gold nanorods on the single-particle level. Adv. Funct. Mater. **22**(4), 702–708 (2012)
35. X. Zhang, N. Tuzzio, M. Tehranipoor, Identification of recovered ics using fingerprints from a light-weight on-chip sensor, in *Proceedings IEEE-ACM Design Automation Conference*, June 2012, pp. 703–708
36. U. Guin, X. Zhang, D. Forte, M, Tehranipoor, Low-cost on-chip structures for combating die and IC recycling, in *Proceedings of ACM/IEEE on Design Automation Conference*, 2014
37. C. Hu, S. C.Tam, F.-C. Hsu, P.-K. Ko, T.-Y. Chan, K. Terrill, Hot-Electron-Induced MOSFET Degradation - Model, Monitor, and Improvement. IEEE J. Solid State Circuits **20**(1), 295–305 (1985)
38. B. Tudor, J. Wang, Z. Chen, R. Tan, W. Liu, F. Lee, An accurate MOSFET aging model for 28 nm integrated circuit simulation. Microelectron. Reliab. **52**(8), 1565–1570 (2012)
39. J. Chen, S. Wang, M. Tehranipoor, Efficient selection and analysis of critical-reliability paths and gates, in *Proc. GLSVLSI*, 2012, pp. 45–50
40. Synopsys, 90nm Generic Library, www.synopsys.com/COMMUNITY/UNIVERSITY PROGRAM/Pages/Library.aspx
41. X. Zhang, M. Tehranipoor, Design of on-chip light-weight sensors for effective detection of recycled ICs, in *IEEE Transactions on VLSI Systems*, 2013
42. B. Stamme, Anti-fuse memory provides robust, secure NVM option, in *EE Times*, July 2012
43. Technology. [Online]. Available: http://www.sidense.com/technology
44. XPM embedded non-volatile memory (NVM). [Online]. Available: http://www.kilopass.com/products/otp-memory-ip/xpm-otp-nvm/
45. A.T. Appel, Rectangular contact used as a low voltage fuse element. Patent US6 774 457 B2, Aug 2004
46. H.C. Nicolay, Integrated circuit fuse. Patent US 4 272 753, Jun 9 1981
47. B. Kaminska, K. Arabi, I. Bell, P. Goteti, J. Huertas, B. Kim, A. Rueda, M. Soma, Analog and mixed-signal benchmark circuits-first release, in *Proceedings of International Test Conference*, 1997, pp. 183–190

Chapter 10
Hardware IP Watermarking

The persistent trend of semiconductor scaling has caused the IC design process to leap forward and at the same time, be held back. On the one hand, with increasing logic density, we are now able to fit more and more components onto a semiconductor die and create functionally dense System-on-a-Chips (SoCs). On the other hand, system complexity has grown exponentially. In order to create and implement new designs, a much larger amount of time and labor is now necessary than what would have been required a few decades back. Unfortunately, with current market trends and fierce competition, time is a very stringent constraint. In order to optimize the design process and decrease time-to-market, the IC industry has shifted gears to the concept of design reuse. Instead of designing a new SoC and its components from scratch, companies nowadays obtain licenses for various functional blocks and work on integrating them into a complete system. This simplifies the design process, as routine functions can be achieved in a circuit by using these pre-designed blocks and leaves more time for innovation. This practice is most commonly seen in the design of SoCs. For example, SoCs such as the one in Fig. 10.1 and those used in mobile processors could have multiple functional blocks for memory, graphics processing, communication and central processing, with each block coming from separate vendors. These pre-designed blocks are termed as semiconductor intellectual property (IP).

In this chapter, we will present an emerging issue of hardware IP protection. As electronic systems today get more complex, the concept of IP reuse has emerged whereby system designers integrate IPs from numerous IP vendors. With this new model of system design, the problem of IP piracy and theft has arisen. To address this issue, the concept of watermarking, which originated from the domain of multimedia, has been applied to hardware IPs. Watermarking in hardware IPs is the technique of embedding a signature into an IP structure, so that the author or owner of the IP can be verified when the watermark is extracted, i.e. a proof of authorship is established in IPs, which can be used to deter illicit use. The rest of the chapter is organized as follows: we will first describe various types of IPs

M. Tehranipoor et al., *Counterfeit Integrated Circuits*,
DOI 10.1007/978-3-319-11824-6_10

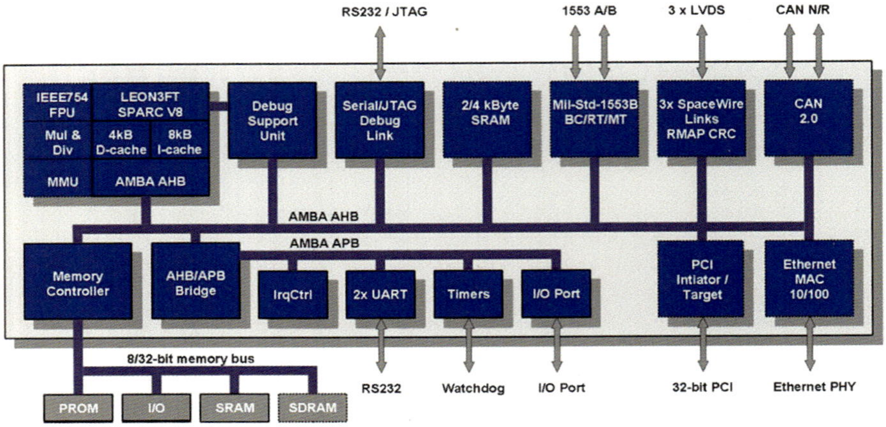

Fig. 10.1 LEON3FT-RTAX SoC processor [1]

and their vulnerabilities. We will then present the concept of watermarking for providing proof of authorship in IP. We will then identify various approaches to integrating watermarks in different IPs while also focusing on the advantages, issues and principles associated with each watermarking technique.

10.1 Intellectual Property (IP)

In the new business model for the semiconductor industry, the design process starts with IP vendors who create reusable logic blocks and standard cells. IP vendors specialize in producing their IP blocks as a result of which these IP blocks are thoroughly tried and tested, in addition to being designed for ease of integration into multiple systems (plug-and-play). System integrators then obtain the license for these IP blocks and design novel architectures for application specific integrated circuits (ASICs), field programmable gate array (FPGAs) or SoCs by combining numerous IP blocks obtained from several different vendors. The final IC design (in the case of ASICs) is then sent to a foundry for fabrication.

Semiconductor IP can be of various types. Broadly, they are classified into three different categories [2].

- **Soft IPs** take the form of RTL (register-transfer level) abstractions. Since they are HDL (or a similar high-level abstraction), they are digital IPs which are process-invariant and can be used to synthesize gate-level information. Soft IPs are flexible and can be easily ported from one system to another. On the downside, soft IPs provide little to no information on their timing and power consumption parameters.

- **Hard IPs** are commonly in the form of Graphical Database System II (GDSII) files. These IPs are fixed in layout and have predictable performance in terms of power, area and timing. They are usually offered by foundries and are tested in silicon to ensure correctness. They come with the disadvantage of being very rigid in design (confined to one process technology) and lack portability. Hard IPs are more common in analog and mixed-signal applications.
- **Firm IPs** are a sort of compromise between soft IPs and hard IPs. They typically come in the form of fully placed netlists and are more predictable than soft IPs as they provide a gate-level description of the IPs. At the same time, they can be easily ported to various process technologies and can be optimized to meet designer needs.

10.2 IP Reuse and IP Piracy

The semiconductor industry has seen large improvements in productivity owing to the principle of IP reuse. IP vendors are constantly working towards improving their IP and making them more flexible for use in multiple designs. The design community uses these IPs and tests them in new SoCs. A recent survey conducted among leading IC design houses showed that as much as 68 % of a silicon die today contains reused IP [3]. The concept of open-source IPs is also on the rise, where designers work continuously to improve IPs and make them freely accessible to the public. As a result, both vendors and designers contribute to increased productivity in the IC industry. But at the same time, with the increased reuse of IPs, IP piracy has become a grave issue. IPs are the intellectual property of their designer. As with any product, IPs are subject to copyrights, patents, and trademarks. IP piracy leads to the misuse and often, unauthorized use of IPs. Problems could be in various forms:

- Claiming someone else's IP as your own and/or reselling it
- Using IPs beyond the scope of their license, such as open-source IPs being used for commercial purposes
- Not giving an IP designer credit where it is due

A more sophisticated and grave issue in IP piracy comes in the form of **reverse engineering**. Reverse engineering is the process of extracting information from a product. Ethically, there are a lot of gray areas in the question of reverse engineering semiconductor IP. But as set forth by law [4], reverse engineering is legal to a certain extent; it is legally permitted to reverse engineer an IP or a semiconductor design for the purpose of teaching, analyzing or evaluating concepts and techniques. Companies often reverse engineer the product of their competitors in order to understand the advancements made or techniques applied. Techniques could involve grinding away layer-by-layer of the IC in question, monitoring I/O relationships and performing circuit extraction or process analysis [5]. The problem is not reverse-engineering in itself. The act of copying the semiconductor design after reverse engineering and using it for commercial benefit while violating patents, copyrights and trademarks is unethical and in most cases, punishable by law [4]. The same scenario applies to IPs. Entities who find ways to reverse-engineer an IP could potentially cause IP piracy, where an IP is used while violating the rights of the IP owner. Piracy could lead to cloning, which makes it possible for parties to make copies of IPs and sell the IP or an IP-integrated product for financial gain, without any licensing or knowledge of the IP authors. Moreover, for any design process, reverse engineering a product and marketing it could be exponentially easier and more financially lucrative than designing a new product altogether. This provides huge incentives for unethical parties to commit IP piracy.

10.3 Approaches to Secure IP

Several measures have been proposed over the years to protect semiconductor IP.

- **Self-Destruction:** For military applications, IPs are often integrated into ICs with chemical destruction mechanisms, which are triggered if any kind of tampering or reverse-engineering is attempted [6].
- **Obfuscation:** The IP structure and functionality is concealed [7] by methods such as changing the structure/content of an HDL component so as to prevent ease of reverse-engineering [8] or, by embedding a finite state machine in the IP which does not allow normal IP behavior unless activated by a valid license key [9].
- **Periodic licensing:** Timers and license controllers are embedded permanently into IPs which keep track of the license period for a user. Once the license period expires, the IP is rendered useless [10].
- **Encryption:** FPGA bitstreams are meticulously encrypted by using a strong encryption algorithm and the encryption key is kept safe in order to prevent reverse engineering of the FPGA IP. For FPGA units such as the Xilinx Virtex-6, encryption/decryption of the bit stream is performed with a 128-bit AES (Advanced Encryption Standard) private key that is stored in a one-time programmable non-volatile memory. This makes sure that the bitstream is not intercepted during transmission and used for reverse-engineering purposes [11].

In terms of legal measures, hardware IPs are most often protected by patents, trademarks, copyrights and trade secrets. These measures deter IP piracy by imposing legal punishment on those who violate IP integrity. Nonetheless, in order to legitimately claim IP theft, IP authors need some sort of watermarking strategy, in order to identify themselves with their design. Further, to prevent ease of direct copying/cloning of IPs, a signature needs to be embedded into an IP design so that IPs can be identified by their unique designers. This brings up the concept of hardware IP watermarking.

10.4 Hardware Watermarking

Hardware watermarking is the process by which a unique fingerprint is embedded into a hardware IP. Watermarking is not a new concept and has been widely adopted in the field of digital media. Images, audio, and videos contain watermarks that prove the authorship of the media. Digital media watermarks can be visible or invisible; the visible ones are logos/signatures that appear on the media itself and invisible watermarks are those that have been incorporated into the media item and are invisible due to imperfections in the human audio-visual system. A distinct feature (or flaw) of digital media watermarks are that they are invasive; regardless of being visible or invisible, the digital data is modified in order to incorporate the watermark, in some way or the other. For the case of digital media, this is acceptable in most circumstances. But applying the same concept of watermarking to hardware IPs is challenging because a watermark should not alter the functionality of the IP. For example, one cannot simply incorporate random interconnects into a layout or add extra lines of code to an HDL file to create a watermark in a hardware IP. Doing this changes the functionality and correctness of the IP and in most cases, such as the adding lines to a HDL file, the supposed watermark gets removed/ignored by HDL compilers during processing, rendering the watermark useless. Considering all these factors, there are multiple requirements for designing a hardware IP watermark that can effectively provide proof of authorship. The now defunct Virtual Socket Interface Alliance (VSIA) pointed out key features [12] that are desired in IP watermarking. An extract of those criteria are listed below.

- **Functional correctness**: The watermark must not alter the core functionality of the IP.
- **Minimal overhead**: The watermark should not affect the IP in terms of performance. Parameters such as power consumption, delay and speed of operation should be negligibly affected.
- **Persistence**: The watermark should be difficult to remove or copy. It should be at least as hard to completely reverse engineer the IP as it is to modify/tamper the watermark. Also, the watermark must be tamper-resistant so that it may not be transformed into a different watermark, in which case proof of authorship can't be proven and third parties may claim ownership of the IP.
- **Invisibility**: The watermark should not be readily detectable by third parties. It should be concealed and only parties with ownership to the IPs should be able to reveal them.
- **Proof of Authorship**: No third party should be able to claim the watermark by chance, i.e. the probability of occurrence of a particular watermark on a non-watermarked IP core must be very low (refer to Sect. 5.1 for a mathematical characterization for P_c, the probabilistic proof of authorship).

In most IP watermarking techniques, a signature is acquired from the author of an IP. This signature, which could be a string or a pattern, is usually converted to a binary sequence, where each bit in the sequence becomes one bit of the watermark.

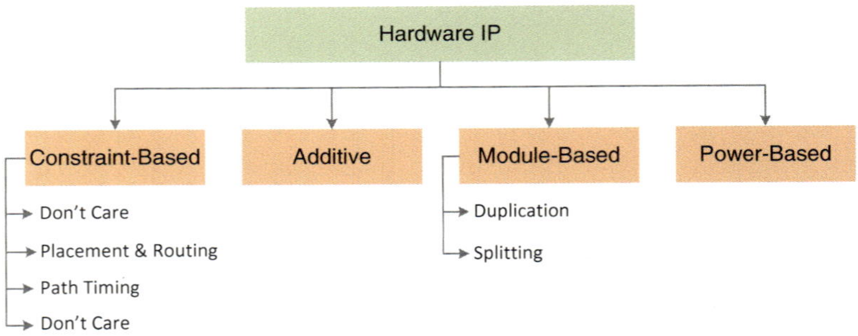

Fig. 10.2 IP watermarking strategies

The signature is also encrypted using a key. The signature is then converted to a watermark by applying various constraints, i.e. watermarking requirements. The constraints may be applied during the design process of an IP or it may be added on to the final IP design as a post-production process. In order to claim authorship when required, the watermark is extracted and verified from the design. Probabilistically, this can only be done by the author/owner as they are in possession of the signature and the key used to decrypt the watermark.

A plethora of techniques exist for watermarking various types of IPs. Figure 10.2 shows a generalized taxonomy for IP watermarking techniques.

- Constraint-based watermarking: The watermark is embedded as a constraint along with the design-constraints of the IP, leading to a combined solution for the original IP design problem and the watermark inclusion problem. Proposed constraints include:
  - Don't care-based watermarking: Don't care conditions that do not affect the functionality of the IP core are imposed as constraints to create a watermark in an IP design.
  - Placement and Routing: Layout IPs are watermarked with specific placement and routing strategies that embed a signature to indicate authorship.
  - Path timing: Path delay constraints are broken down into sub-path delay constraints to indicate an author-specific signature.
  - Cache-line coloring: Signature-specific nodes are assigned to graph coloring problems frequently used in cache-memory organization.
- Additive watermarking: FPGA IPs are watermarked by adding the watermark to the functional core for the IP.
- Module-based watermarking: IPs in the form of hardware description languages (HDL) are watermarked by either duplicating frequently used blocks of HDL code (module duplication) or by dividing an HDL module into sub-modules (module splitting), while maintaining the same functionality as a non-watermarked HDL IP.
- Power-based watermarking: A signature is extracted from the power consumption pattern of an FPGA unit.

10.4.1 Constraint-Based Watermarking

During the process of IP design, there are certain functional constraints that are taken into account. Criteria such as delay and power consumption dictate the IP design process. The concept of constraint-based watermarking, first introduced by Kahng et al. [13] is to add new "watermarking constraints" into the IP design process. In this process, the author of an IP creates a signature and uses a secret key to convert the signature into a set of constraints, making sure that these constraints do not conflict with the original IP design constraints. The solution to an IP design problem then becomes one that not only satisfies the functional constraints but also the added watermark constraints.

If designers need to verify their authorship, they would use their private key along with their signature to generate the overall solution. In terms of verification, the solution space can be visualized as shown in Fig. 10.3. The solution space of the IP design problem is very big, i.e., there are numerous solutions available. But at the same time, consider the limited amount of solutions for the watermark IP problem. Combine the number of solutions that satisfy both the IP design and watermark IP problem and it can be clearly seen that the number of solutions are very few. Thus, the probability that any constraint-based problem is solved by solving just the original problem becomes very low. Thus, the solution must have come from the combined IP design + watermark design solution. This provides a probabilistic proof of authorship [14]. Mathematically, this is expressed as [13]:

$$P_c = P(X \leq b) = \sum_{i=0}^{b} \left[(C!/(C-i)! * i!) * (p)^{C-i} * (1-p)^i) \right] \tag{10.1}$$

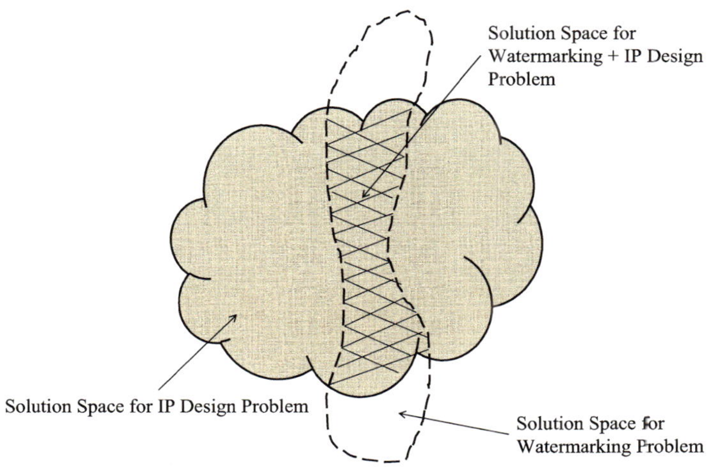

Fig. 10.3 Constraint-based watermarking, adapted from [14]

where, P_c = proof of authorship
p = probability of satisfying one random constraint by coincidence
C = number of imposed constraints
b = number of constraints unsatisfied
X = random variable that represents how many of the imposed constraints (C) were not satisfied.

In this expression, P_c is the probability that a non-watermarked solution carries a particular watermark by coincidence. When designing constraint-based watermarking strategies, it is obvious that we need to keep P_c as low as possible so that the final solution for the IP design + watermark problem can only be satisfied by the author. An appropriate P_c could be in the range of 10^{-30} indicating that it is not feasible (time and effort-wise) for anyone to copy the watermark [15]. Thus, constraint-based watermarking ensures a reliable approach to IP watermarking, where the chance that anyone could randomly guess the correct solution without the signature, key etc. is very negligible.

The boolean satisfiability problem (SAT) illustrates the concept of constraint-based IP watermarking [13]. The SAT problem seeks to find whether there exists a solution (conditions for a logical TRUE) for a given boolean expression. Let us take a finite set of variables $U = \{u_1, u_2\}$ and let $C = \{\{\bar{u_1}, u_2\}, \{u_1, \bar{u_2}\}, \{u_1, u_2\}\}$ be the set of some clauses, i.e., disjunction of variables from the set that need to be satisfied. Enumerating the solution set shows that there are three solutions that satisfy this SAT problem ($\{u_1 = T, u_2 = F\}$ or $\{u_1 = F, u_2 = T\}$ or $\{u_1 = u_2 = T\}$) . Now, let us impose an additional constraint into this set of solutions. Take the clause $\{\bar{u_1}\}$, which could be a kind of signature identifying the author and add it to the set of clauses. Now, the solution set decreases to 2 ($\{u_1 = F, u_2 = F\}$ or $\{u_1 = F, u_2 = T\}$). This simple example of how adding a constraint decreases the solution space shows us how the chance of anyone randomly guessing the correct solution is dramatically decreased, illustrating that anyone who can solve the overall solution (functional + watermark constraint) has authorship rights to the IP.

10.4.1.1 Don't Care Condition-Based Watermarking

Don't care condition-based watermarking falls within the domain of constraint-based watermarking as well. In this technique, one makes use of truth tables, the functional backbone of any digital logic. In any truth table, there might be instances where the designer does not really care what the output(s) is/are for certain pair of inputs. These input combinations are termed as 'don't care' conditions. For IP watermarking, don't care conditions can be used as functional blocks which force the output of the IP. An example would be a boolean expression such as $f(a, b, c, d) = \bar{a}b\bar{c} + \bar{a}bd + b\bar{c}d$. In order to embed 1 bit of a watermark signature (e.g., logic 1), input combinations that force the output to go high are added as don't care conditions to the boolean expression. At the same time, to assert a logic 0 for the signature, the same input combinations causing a don't care condition are removed. In the given expression, to assert a logic 1, the don't care term $\bar{a}\bar{b}\bar{c}\bar{d}$ is added. The term $\bar{a}\bar{b}\bar{c}\bar{d}$ equals logic 1 if and only if all of the inputs $a, b, c\ and\ d$ are all low. For any other combination of inputs, the $\bar{a}\bar{b}\bar{c}\bar{d}$ term would be a don't care condition in the overall boolean expression [14].

10.4.1.2 Placement and Routing-Based Watermarking

Placement and routing watermarking techniques are applied to hard IPs, where constraints are placed in the physical design level. **Row-based placement techniques** are applied as a post-processing step where a signature is encoded as a specific parity (odd/even) of a cell row where standard cells must be placed [16]. In the physical floorplan of the IP design, legal cell placement locations are arranged in a row. First, a message is taken from the IP designer which is then converted into row-parity constraints for a subset of the cells of the design. The selected cells are then arranged as per the generated row constraints using pair-swap or replacement and finally, routing is performed to generate a final watermarked design. Thus, the final design would contain cells that are arranged specifically, according to the placement constraints generated by the signature and the watermark would be concealed within the grid abstraction of the layout. Placement-based watermarking deters tampering, as a considerable number of swap operations for the cells would be required for an attacker to damage the watermark, by which time the IP itself would have been rendered useless.

Similarly, **routing-based watermarking** can also be used to impose constraints on the way nets are routed in a physical design [16]. A signature from the IP designer is obtained and converted into a list of "watermark nets" which are then chosen from the set of all nets in the IP design. Each watermark net is assigned an unusually low cost in terms of wrong-way wiring, i.e., the length of acceptable wrong-way routing is limited. This distinguishes the watermark nets from other nets in the design. The watermarked nets are then specified into the routing protocol of the EDA tool. In order to identify the watermark nets, a ratio is then computed by dividing the total wire length (WL_{tot}) and wrong-way wirelength (WL_{way}) of each watermark net. If this ratio is less than an assigned threshold, the net can be classified as a watermark net. The advantage of using a routing-based approach is that any attempt at tampering with the routing of the nets will lead to a quicker degradation of the IP solution quality than the watermark itself. In place of using a cost-based routing approach, other constraints such as wire width, spacing and topology could be used to incorporate a watermark into the layout of an IP. In another approach, a signature could be converted into a bit-stream watermark. The bits from the watermark are embedded into a group of linearly ordered nets [17]. If a single net index maps to a '1' in the bit-stream, the net is rerouted with an even number of bends. For a '0', the net is rerouted with an odd number of nets. With a system of odd and even number of bends in the routing, a watermark could be embedded into a layout (see Fig. 10.4).

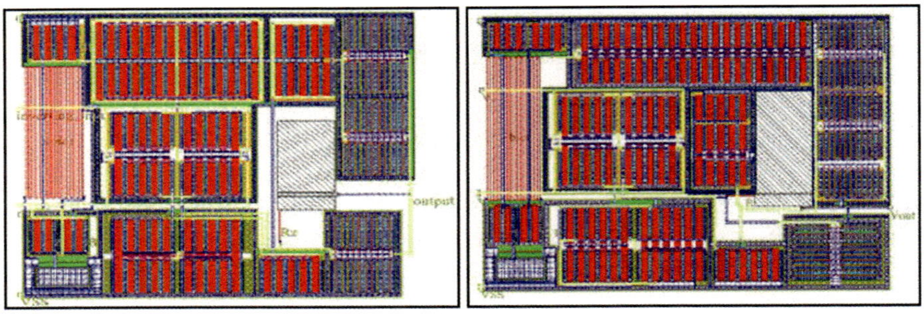

Fig. 10.4 Non-watermarked routing (*left*) and watermarked routing (*right*) for a two-stage Miller operational amplifier [17]

However, routing-based strategies could cause concerns as nets could be forced into inefficient routes, leading to crosstalk between wires and ultimately, performance degradation.

10.4.1.3 Path Timing Constraint-Based Watermarking

In this approach, path-based timing is considered as a constraint in the design of the IP. Numerous parameters such as interconnect RC characteristics and floor-planning could all qualify as timing constraints. Given a path timing constraint, this approach suggests that the path timing constraint be replaced by subpath timing constraints [13]. For example, take the path timing constraint for cells in a design $t(C_1 \rightarrow C_2 \rightarrow C_3 \rightarrow \ldots \rightarrow C_{10}) \leq 50ns$, where C_i refers to each cell in the design. This timing constraint can now be divided into two constraints $t(C_1 \rightarrow \ldots \rightarrow C_5) \leq 20ns$ and $t(C_5 \rightarrow \ldots \rightarrow C_{10}) \leq 30ns$. Dividing the path timing constraint makes sure that the probability of randomly satisfying the original path constraint and consequently, satisfying the subpath constraints is very low; this is a feature of constraint-based watermarking.

10.4.1.4 Cache-Line Coloring

In a processor architecture, instruction and data caches occupy a large portion of silicon area and are critical components for efficient system performance in terms of timing and power consumption. In order to minimize the cache-miss ratio in the cache memory structure, graph coloring, the technique of coloring the vertices of a graph so that no two adjacent vertices have the same color, is employed in order to maximize the total number of pages cached by a processor. By ‘coloring’ physical memory addresses, we can make sure that adjacent virtual memory spaces do not map to the same position in the main cache memory, alleviating the problem

Fig. 10.5 Watermarking a graph coloring problem [18]

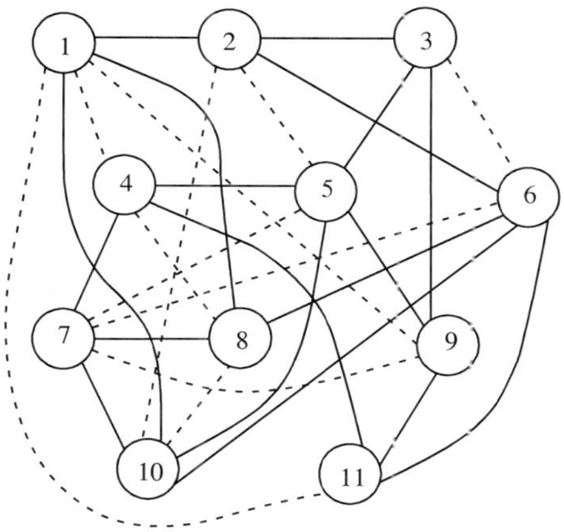

of cache conflict. In its simplest implementation, the problem of mapping code to cache could be represented as a control data-flow graph which could then be treated as a graph coloring problem.

In order to watermark such designs, a constraint-based watermarking approach could be employed. The designer's signature could be converted into a set of constraints, such as a binary string, by using a private key. The constraints could then be assigned as additional nodes in the graph-coloring problem. As a result, the final code-to-cache mapping would contain the solution as well as the signature of the author [18]. Figure 10.5 shows an example of watermarking a cache line coloring problem. The graph shown is embedded with the signature $1998_{10} = 11111001110_2$. Each bit of the signature corresponds to the dotted lines/edges in the graph. Retrieving the signature would involve reconstructing the binary string (the constraint) and then recreating the graph with the extra edges. If the coloring of the watermarked graph is a valid coloring of the recreated graph, the signature can be verified. Unfortunately, such a watermarking technique is prone to attacks by modifying the watermarking coloring and extracting the signature of the author, which makes it possible for anyone to claim authorship of the coloring solution [19].

10.4.2 Additive Watermarking

Additive watermarking techniques embed the signature into the functional core of an IP design but do not modify the functions of the IP core like constraint-based watermarking techniques [20]. They are mostly incorporated into the unused portions of an IP design (such as unused lookup tables and ports) and can be incorporated during pre-processing or post-processing. The disadvantage of such a technique is that since the watermark is not a part of the functional design, the watermark may be removed without affecting the IP core functionality. Nonetheless, various techniques are applied to mask the watermarks and make them seem as part of the IP core's function. Additive watermarking techniques are popular in FPGA IP watermarking.

10.4.2.1 Watermarking in Physical Level FPGA Design

Watermarking can be applied to the reconfigurable logic layout of an FPGA. FPGAs are made of an array of configurable logic blocks (CLBs) that consist of flip flops and multiple look up tables (LUTs). In every FPGA design, not all CLBs are used. For every CLB that is unused, there are several lookup tables that are also freely available. Each LUT is capable of holding one bit or several bits of information and with a lot of bits available, an author's signature could be integrated into the FPGA physical design [21]. In order to physically achieve this, a netlist and a signature is first read in. Using standard vendor tools, the unused netlist is routed. We then proceed if the unused netlist has enough resources to incorporate the signature. The signature is then processed using encryption and error-correcting codes in order to make sure the signature's integrity can be maintained even if it is tampered with. The processed signature is then coded to the available LUTs using a secure hash function. While the process is fairly straight-forward, implanting watermarks in the physical level of an FPGA design causes overhead in the form of area and timing. The watermark LUTs may affect normal routing of the cells and of course, since additional components are now part of the FPGA design, the footprint is obviously going to be bigger. In terms of timing, a LUT which is a part of the watermark may occasionally be placed in timing-critical paths, which may result in undesired delays.

Another approach for watermarking an FPGA design at the physical level is to insert watermarks into the control bits for CLB outputs [22]. Usually, outputs of CLBs are controlled by multiplexer units. FPGA units such as the Xilinx 4000 series have CLBs with four outputs (see Fig. 10.6). The two outputs X and Y are outputs from combinational designs while YQ and XQ are outputs from sequential designs. The combinational outputs are controlled by two 2–1 multiplexers, each with one control bit. The sequential outputs have two control bits for the 4–1 multiplexers and one control bit each for the three 2–1 multiplexers. If the CLB is unused, a total of $(2 \times 1) + (2 \times 2) + (6 \times 1) = 12$ control bits can be used to encode a signature. The process of encoding a signature is straightforward. FPGA design tools are used to scan the FPGA IP Design and find CLB outputs that are not connected to any external CLB interconnects. Bits are then sequentially slotted in place of the multiplexer control bits. Extracting the watermark bits is an identical process. Again, the FPGA design tool scans the IP architecture, finds any CLB outputs that are unused and extracts the watermark from the control bits of the multiplexer. Since this strategy is purely post-processing, no performance or area overhead is incurred. The watermark size will only be limited by the number

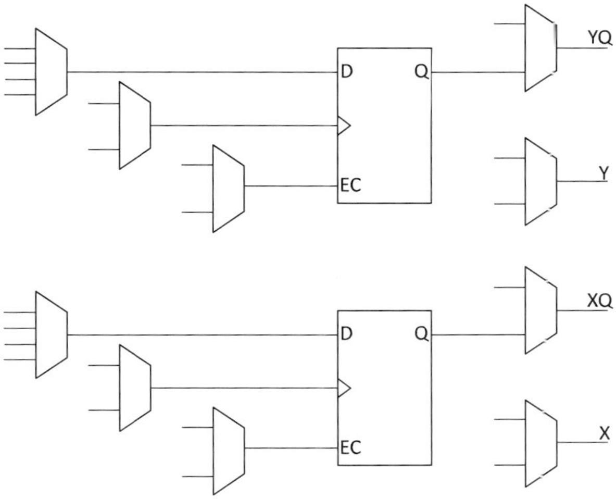

Fig. 10.6 Control bits for CLB outputs [22]

of unused CLB outputs available, which is usually large in any FPGA design. Nonetheless, such a watermarking scheme is prone to reverse-engineering attacks as the watermark itself is a non-functional part of the IP design. Control-bit watermark security is largely ensured by the confidentiality of the FPGA bitstream, which inhibits reverse-engineering.

10.4.3 Module-Based Watermarking

Module-based watermarking mainly concerns the protection of soft IP, in the form of HDL codes. From a security standpoint, hard IPs such as GDSII files are the safest as they are very hard to reverse-engineer or tamper. Unfortunately, hard IPs are stringently optimized for performance and are silicon process-specific, making them inflexible. Soft IPs, on the other hand, allow a greater level of flexibility and in some cases, can be modified by IP users for further optimization. This has led to the widespread use of soft IP and at the same time, raised concerns regarding soft IP security. Hard IP watermarking measures such as placement/routing and path timing are inapplicable to soft IPs. On top of that, traditional source-code watermarking measures such as obfuscation cannot be used for HDL codes. Obfuscation involves making a program unintelligible yet functional. The industry's push for standardizing HDL to improve IP reusability defeats this strategy. Thus, for watermarking in soft IPs, a slightly different approach is required.

10.4.3.1 Module Duplication

In Verilog code, more often than not, we find basic functional modules that are called on multiple times by other modules either in the same or a higher hierarchy. These modules, that are repeatedly instantiated, can be coded in different ways to achieve the same function. This is the basis for module duplication. In the case that a module contains don't care conditions, the values for these don't care conditions can be assigned different values to create multiple modules from the same module. In the absence of don't care conditions, we can simply implement the module in a different manner. This is to make sure the synthesis tool does not delete the duplicated modules during optimization. For example, we can consider a pattern detector that detects the binary sequence '1101'. The sequence detection can be depicted by a state transition diagram as shown in Fig. 10.7. A Verilog module of the pattern detector could be designed in two ways, by either using a finite state machine (Code A) or a shift register (Code B). Both versions of the module would achieve the same goal: output a '1' if the binary sequence '1101' occurs, otherwise output a '0'. A separate module could then be built (Code C), which could output a 1 if the FSM module is picked and output a '0' if the shift register module is picked. Then, a signature could be made from the number of times each modules is called. Module duplication can be implemented into existing HDL codes and for anyone attempting to reverse engineer the HDL IP, it becomes extremely difficult to distinguish the duplicate modules as the modules have already made it past the synthesis tool. On the contrary, duplicating modules several times causes a significant area and performance overhead in the final IP design [23].

```
module detector_0 (clk, reset, dataIn, out);
   input clk, reset, dataIn;
      output out;
      reg out;
      reg  [1:0] currentState, nextState;
        always @(dataIn or currentState) begin
       case (currentState)
        2'b00: begin
           nextState = (dataIn == 1) ? 2'b01 : 2'b00;
           out = 0;end
        2'b01: begin
               nextState = (dataIn == 1) ? 2'b10 : 2'b00;
           out = 0;end
        2'b10: begin
           nextState = (dataIn == 0) ? 2'b11 : 2'b10;
```

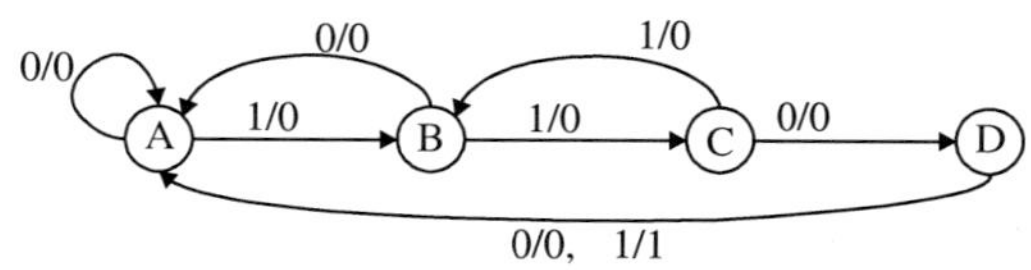

Fig. 10.7 Pattern detector for the binary sequence '1101' [14]

```
                out = 0;end
         2'b11: begin
             nextState = 2'b00;
             out = (dataIn == 1);end
       endcase
    end
    always@(posedge clk) begin
       if(~reset) begin
                currentState <= 2'b00;
             out <= 0;
       end
       else currentState <= nextState;
    end
endmodule
```

Code A: Finite State Machine Implementation of Pattern Detector [23]

```
module detector 1 (clk,reset,dataIn,out);
    input dataIn,clk,reset;
    output out;
    reg out;
    reg [3:0] pattern;
    always@(posedge clk) begin
        if( reset) begin
            pattern = 0;
            out = 0;end
        else begin
            pattern[0]=pattern[1];
            pattern[1]=pattern[2];
            pattern[2]=pattern[3];
            pattern[3]=dataIn;
            if(pattern==4'b1101) out=1;
            else out=0;
        end
    end
endmodule
```

Code B: Shift Register Implementation of Pattern Detector [23]

```
module P;
    reg clk, reset;
    reg data1,data2,data3;
    wire out1, out2, out3;
    detector_0 d1(clk, reset, data1, out1);// signature bit 0
    detector_1 d2(clk, reset, data2, out2);// signature bit 1
    detector_0 d3(clk, reset, data3, out3);// signature bit 0
    . . .    . . .    . . .
endmodule
```

Code C: Module Detector [23]

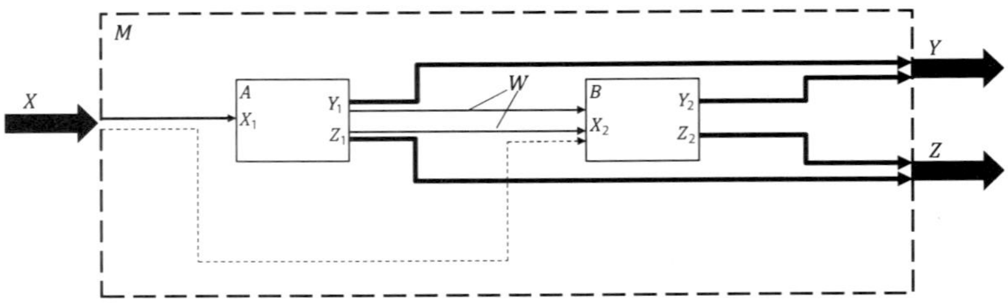

Fig. 10.8 Module splitting [23]

10.4.3.2 Module Splitting

For HDL IPs that contain fairly large modules, the large modules can be broken up into several smaller ones [23]. Figure 10.8 shows how a module $M(X, Y, Z)$ can be subdivided into modules $A(X_1, Y_1, Z_1)$ and $B(X_2, Y_2, Z_2)$. Module A first takes in the input X_1 and produces the watermarking outputs W along with part of the output (Y_1) and part of the test output (Z_1). Module B receives the input X, the watermarking outputs and at the output of the entire module M, the actual outputs Y and Z are reconstructed by a set union. The output produced by A and B combined is the exact same as the output produced by module M, which preserves functional correctness. In order to retrieve the watermark, the input is fed into module A and the watermarking signal W is observed. Since the watermark is part of the functional design of the IP, synthesis tools will not remove the watermark and the watermark will not incur as much overhead as a module duplicating IP would. A major drawback to this approach, however, is the increase in design complexity.

10.4.4 Power-Based Watermarking

In this approach, a signature is extracted from the power consumption pattern of an FPGA unit and the watermark is detected at the power supply pins of an FPGA unit [24]. In an FPGA unit, consumed power can be of two varieties: static and dynamic. Static power consumption comes in the form of leakage current from CMOS transistors while dynamic power consumption comes from the switching of transistors, where capacitance is constantly reloaded and short circuit current occurs along with the operational clock edge. With such transient activities, we can also notice that the core voltage of an FPGA keeps fluctuating with the consequent breakdowns and overshoots. Due to this, a voltage versus time plot of the voltage supply pin on the FPGA shows the clock frequency or integer divisions of the clock frequency. In order to embed a watermark using power consumption, a power-draining component, such as an additional shift register can be embedded into the FPGA architecture. The shift register can be clocked separately by a combinatorial logic or a separate clock, which runs at a frequency that is different from the operating frequency of the FPGA. Spectral analysis of the power supply pins in the FPGA would then show two peaks, one at the operating frequency and one at the unique frequency of the separate clocking logic, which appears whenever the power-intensive component is clocked and can be used as a watermark after repeated sampling and decoding. However, jitter in the clock from the combinatorial logic could possibly make the watermark frequency hard to detect.

Alternatively, amplitude could be used as a parameter for power-based watermarking. A power-draining component such as a shift register is used to generate a binary pattern, but with the operational clock as the clocking source. An additional control logic based on a signature is implemented so that whenever the logic is '1', the shift register outputs a bit otherwise, it stalls. A voltage profile over time is then constructed by monitoring the voltage supply pins on the FPGA. The watermark would then be observed as a series of high and low amplitudes on the waveform (Fig. 10.9).

Both of these approaches represent non-invasive methods of watermarking, where bitfiles and extraneous routing are not necessary. However, a large overhead is incurred in terms of power consumption. Further, power analysis-based watermarking techniques are subject to a number of attacks [25]. Depending on the level of abstraction that an attacker has access to, it may be possible to reverse engineer an FPGA IP and completely remove the power-based watermarking circuit, such as

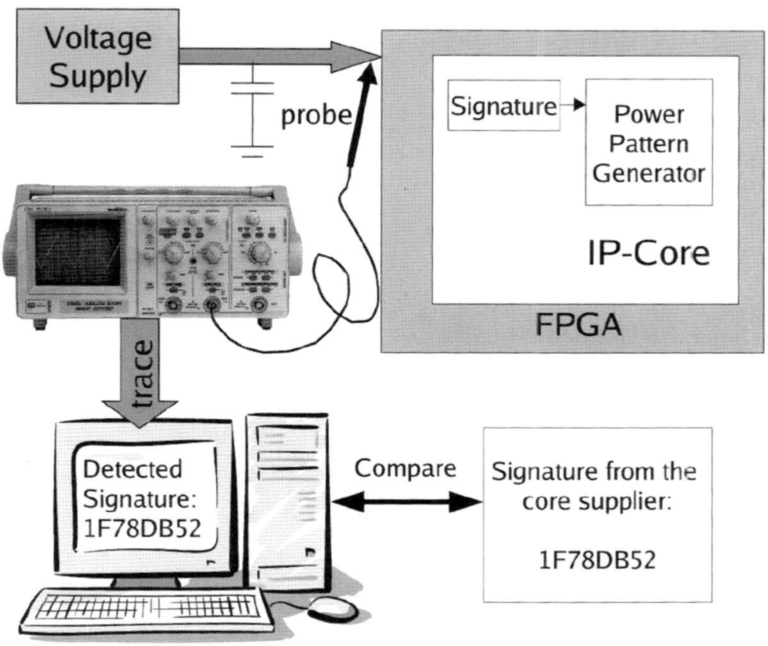

Fig. 10.9 Watermark verification via power analysis [24]

the shift register described above. Thus, the watermark will only be as secure as the FPGA IP itself. Another concern for the power-based watermarking approach is the signal-to-noise ratio (SNR). An attacker may be able to inject additional components into the FPGA IP in order to lower the SNR, which makes it difficult for a power-based watermark to be identified.

10.5 Summary

In this chapter, we presented the concept of IP reuse and brought into perspective the need for protecting the integrity of hardware IPs. We also discussed several different methods for watermarking hardware IPs to provide proof of authorship. Although most of the watermarking strategies are provably secure, several types of attacks could still affect the watermark. For constraint-based watermarking, the watermark would only be as safe as the encryption key used to convert the signature into a set of constraints. Once someone receives the private key for a constraint-based watermarked design, he or she could easily forge the IP design. FPGA watermarking strategies such as using unused CLBs could be removed by tracing unused pins on the FPGA to the unused CLBs and possibly zeroing the multiplexer control bits on the CLBs to tamper/remove the watermark. Another possible threat are ghost signatures, where a third party could declare that either someone's watermark is not in the IP design where it is supposed to be, or they could declare that their watermark is in the design when it actually isn't there. This could be done by working out the input pattern from the solution set of the watermarking problem. Thus, additional security measures are needed for pre-existing watermarking techniques. To achieve this, watermarks could possibly be distributed in small sizes all over an IP, as opposed to just being in one area. In addition, watermarks could be built hierarchically by integrating watermarks in each step of the design process (from HDL all the way to silicon). Parity checking could also be used to ensure that watermarks have not been tampered. Constraint-based watermarking should employ techniques to reduce the possibility that someone could guess or extrapolate to a ghost signature. Regardless, one needs to realize that no matter how tamper-proof or secure watermarks are and how much they can contribute to proving authorship, they have a fundamental limitation. Watermarks are "passive" in that they can only be used for proof of authorship/ownership during the litigation process. They cannot be used to "actively" prevent reverse-engineering or cloning of an IP in the first place, by which time critical information regarding an IP and its confidentiality might have already been compromised. Active methods of IP protection will be discussed in Chap. 11.

References

1. A.B. Aeroflex Gaisler, *LEON3-FT SPARC V8 Processor*, Data Sheet and User's Manual, Aeroflex Gaisler AB Std., Rev. Version 1.9, January 2013
2. A.T. Abdel-Hamid, S. Tahar, E.M. Aboulhamid, IP watermarking techniques: survey and comparison, in *Proceedings. The 3rd IEEE International Workshop on System-on-Chip for Real-Time Applications, 2003* (IEEE, 2003), pp. 60–65
3. S. Sikand, IP Reuse – Design and Verification Report 2013, Design Reuse, Verification Reuse and Dependency Management, IC Manage, Inc., Tech. Rep., 2013
4. B. Shakya *Protection of Semiconductor Chip Products*, 17 USC, 901–914 (1984), http://copyright.gov/title17/92chap9.htm
5. R. Torrance, D. James, Reverse engineering in the semiconductor industry, in *IEEE. Custom Integrated Circuits Conference, 2007. CICC '07*, Sept 2007, pp. 429–436
6. R.N. Das, V.R. Markovich, J.J. McNamara Jr, M.D. Poliks, Anti-tamper microchip package based on thermal nanofluids or fluids, Oct. 16 2012, US Patent 8,288,857
7. R. Chakraborty, S. Bhunia, HARPOON: An obfuscation-based SoC design methodology for hardware protection. IEEE Trans. Comput. Aided Des. Integrated Circ. Syst. **28**(10), 1493–1502 (2009)
8. M. Brzozowski, V. Yarmolik, Obfuscation as intellectual rights protection in VHDL language, in *6th International Conference on Computer Information Systems and Industrial Management Applications, 2007. CISIM '07*, June 2007, pp. 337–340
9. R. Chakraborty, S. Bhunia, RTL hardware IP protection using key-based control and data flow obfuscation, in *23rd International Conference on VLSI Design, 2010. VLSID '10*, Jan 2010, pp. 405–410
10. N. Couture, K. Kent, Periodic licensing of FPGA based intellectual property, in *IEEE International Conference on Field Programmable Technology, 2006. FPT 2006*, Dec 2006, pp. 357–360
11. E. Peterson, Developing Tamper Resistant Designs with Xilinx Virtex-6 and 7 Series FPGAs," Xilinx, Tech. Rep. XAPP1084 (v1.3), Oct 2013
12. Virtual Socket Interface Alliance, *Intellectual Property Protection White Paper: Schemes, Alternatives and Discussion Version 1.0*, Virtual Socket Interface Alliance Std., September 2000
13. A. Kahng, J. Lach, W. Mangione-Smith, S. Mantik, I. Markov, M. Potkonjak, P. Tucker, H. Wang, G. Wolfe, Constraint-based watermarking techniques for design IP protection. IEEE Trans. Comput. Aided Des. Integrated Circ. Syst. **20**(10), 1236–1252 (2001)
14. G. Qu, L. Yuan, Secure hardware IPs by digital watermark, in *Introduction to Hardware Security and Trust*, ed. by M. Tehranipoor, C. Wang (Springer, New York, 2012), pp. 123–141
15. E. Charbon, Hierarchical watermarking in IC design, in *Proceedings of the IEEE 1998 Custom Integrated Circuits Conference, 1998*, May 1998, pp. 295–298
16. A.B. Kahng, S. Mantik, I.L. Markov, M. Potkonjak, P. Tucker, H. Wang, G. Wolfe, Robust IP watermarking methodologies for physical design, in *Proceedings of the 35th Annual Design Automation Conference* (ACM, 1998), pp. 782–787
17. N. Narayan, R. Newbould, J. Carothers, J. Rodriguez, W. Holman, IP protection for VLSI designs via watermarking of routes, in *Proceedings of the 14th Annual IEEE International ASIC/SOC Conference, 2001*, 2001, pp. 406–410
18. G. Qu, M. Potkonjak, Analysis of watermarking techniques for graph coloring problem, in *Proceedings of the 1998 IEEE/ACM International Conference on Computer-Aided Design* (ACM, 1998), pp. 190–193
19. T. Van Le, Y. Desmedt, Cryptanalysis of UCLA watermarking schemes for intellectual property protection, in *Information Hiding* (Springer Berlin Heidelberg, 2003), pp. 213–225
20. D. Ziener, J. Teich, Evaluation of watermarking methods for FPGA-based IP-cores, *University of Erlangen-Nuremberg, Department of CS*, vol. 12, 2005

21. J. Lach, W.H. Mangione-Smith, M. Potkonjak, Signature hiding techniques for FPGA intellectual property protection, in *1998 IEEE/ACM International Conference on Computer-Aided Design, 1998. ICCAD 98. Digest of Technical Papers* (IEEE, 1998), pp. 186–189
22. A.B. Kahng, J. Lach, W.H. Mangione-Smith, S. Mantik, I.L. Markov, M. Potkonjak, P. Tucker, H. Wang, G. Wolfe, Watermarking techniques for intellectual property protection, in *Proceedings of the 35th Annual Design Automation Conference*, ser. DAC '98 (ACM, New York, NY, USA, 1998), pp. 776–781. [Online]. Available: http://doi.acm.org/10.1145/277044.277240
23. L. Yuan, P.R. Pari, G. Qu, Soft IP protection: Watermarking HDL codes, in *Information Hiding* (Springer Berlin Heidelberg, 2005), pp. 224–238
24. D. Ziener, J. Teich, FPGA core watermarking based on power signature analysis, in *IEEE International Conference on Field Programmable Technology, 2006. FPT 2006* (IEEE, 2006), pp. 205–212
25. G. Becker, M. Kasper, A. Moradi, C. Paar, Side-channel based watermarks for integrated circuits, in *2010 IEEE International Symposium on Hardware-Oriented Security and Trust (HOST)*, June 2010, pp. 30–35

Chapter 11
Prevention of Unlicensed and Rejected ICs from Untrusted Foundry and Assembly

In Chap. 10, we discussed the emergence of IP reuse and the challenges it has created with respect to IP piracy. Another recent trend is the transition of most semiconductor companies to a fabless business model. In the past, a company would have full control of their product from design to fabrication/assembly. However, the costs associated with modern IC fabrication have become prohibitively expensive. Thus, most semiconductor companies have been forced to outsource manufacturing of their designs to contract foundries. This horizontal business model requires that they share their design with untrusted third parties, which has led to many well documented threats including IC piracy/cloning, IC overproduction, shipping of improperly or insufficiently tested chips, and hardware Trojan insertion [1–8]. The appearance of such chips in the supply chain can be catastrophic for critical applications.

As discussed in Chap. 2, cloned and overproduced chips may not be as thoroughly tested as authentic chips. Critical systems (transportation, defense, etc.) that unknowingly use such parts will be prone to failure and could lead to life-or-death situations. In addition, cloned and overproduced chips can reduce the profits and harm the reputation of the IP owner. Similar issues can arise from out-of-spec/defective parts.

In this chapter, we shall focus on prevention of IC piracy/cloning, overproduction, and sourcing of out-of-spec/defective components. We begin by discussing the issues associated with the fabless business model and then describe the approaches that have been proposed to protect semiconductor companies and IP owners from IP piracy, overproduction, and cloning. In contrast to Chap. 10, many of these methods are "active" in the sense that they modify design functionality in a way that protects the IP/ICs from the above threats rather than simply proving IP ownership.

Finally, we shall discuss Connecticut Secure Split-Test (CSST), a technique which prevents sourcing of out-of-spec/defective ICs by untrusted foundry and assembly in addition to cloned and overproduced ICs. CSST gives control over testing back to the IP owner. In CSST, each chip and its scan chains are locked

M. Tehranipoor et al., *Counterfeit Integrated Circuits*,
DOI 10.1007/978-3-319-11824-6_11

during test and only the IP owner can interpret the locked test results and unlock passing chips. The IP owner will also be able to control the number of ICs unlocked. In this way, CSST can prevent overproduced, defective, and cloned chips from reaching the supply chain.

11.1 Fabless Business Model

Fabrication of an IC requires several complex and sensitive steps. The industry that makes an IC is called a foundry or fab. Fabs require expensive, well-maintained equipment and spaces that are completely uncontaminated to avoid catastrophic yield losses. There is little one can do if a die is fabricated incorrectly. Continued scaling and complexity of integrated circuits (ICs) have significantly increased the manufacturing costs in the semiconductor industry. For example, moving from 32 to 28 nm technology node adds 40 % extra manufacturing cost [4–6, 9]. A new semiconductor fab is expected to exceed US $ 5.0 billion by 2015 with large recurring maintenance costs [4–6, 10]. It has been reported that a fab costs around 50 USD per chip for maintenance only.

Prior to the 1990s, fabs were not as expensive and all semiconductor companies had their own fab. The semiconductor ecosystem has changed due to change in semiconductor economy. As a result, most semiconductor companies have closed their fabs [10] and now outsource the manufacturing in order to reduce the cost per IC. This outsourcing model is called fabless business model. In this model, semiconductor companies purchase excess foundry capacity from other fabs. Since these contract gabs are more often located offshore this simultaneously results in less control over IC fabrication by the IP owner and increases the number of vulnerabilities.

Figure 11.1 shows the entire IC supply chain which includes design, fabrication, assembly, and distribution. The design house creates their IP using a combination of in-house design teams, IP from third party vendors, and third party design tools. Once the design has been synthesized and verified, the IP owner

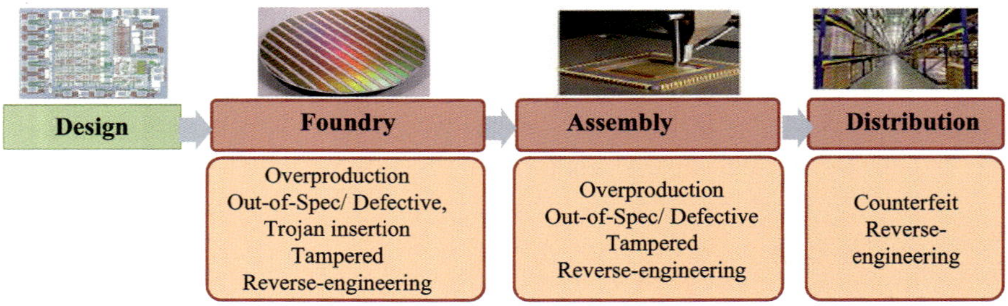

Fig. 11.1 IC supply chain and vulnerability

creates a GDSII layout. The layout, test patterns, and correct test responses will be provided to the contract fab. Typically, the IP owners will have little interaction with their ICs and the fabrication/assembly process from this point forward. The foundry develops a mask which is very expensive in order to manufacture the ICs using photolithography. Once the chips are manufactured, the wafer is tested for correct operation at the manufacturing site or other third party test facilities [6]. Failed die on the wafer are marked with permanent ink and discarded after die are collected from wafer. Dies that pass testing are supposed to be sent to the assembly where they will be packaged and then tested once more. Often, the packaged ICs shall be sent directly from assembly to the market or to the IP owner's vendors. It is the responsibility of the assembly to ship only good ICs to the market in the volume requested by the IP owner.

11.2 Fabless Supply Chain Vulnerabilities

The high cost of IP development puts the parties involved in IC manufacturing and testing in a position where it is possible to profit from exploitation of the IP they have been provided with. A GDSII file contains the whole IP design [4, 6]. Untrusted entities in the fab can tamper with the GDSII. Untrusted fab can also overproduce (i.e., produce more ICs than contracted to) and send out-of-spec/defective ICs into the market (through lax or nonexistent testing). Similar attacks are possible at assembly as well.

The shift to this horizontal business model and use of untrusted third parties has brought major concerns to industry, governments, and consumers. Counterfeit ICs can have a major impact on the security and reliability of critical applications. They are of great concern to government and industry because of the life threatening situations they create and the negative impact they can have on innovation, economic growth, and employment. For the remainder of the chapter, we will focus on the following counterfeit types and proposed mitigations:

- **Overproduced**: In the horizontal business model, any untrusted foundry/assembly that has access to a designer's IP can exceed the agreed volume contact and sell the overproduced ICs on the grey/black market. This attack is economically motivated because the fab can make a large profit without the IP development costs incurred by the IP owner [3–5, 11].
- **Cloned:** A clone is an unauthorized production without a legal IP. Cloning can be accomplished through IP theft, espionage, or reverse engineering. Reverse engineering [12, 13] recovers unavailable specifications of integrated circuits which can be used to reproduce ICs. For example, reverse engineering can be accomplished by studying mask data of the IP. The mask data is then converted to functional level through transistor level netlist and gate level abstraction.
- **Out-of-Spec/Defective**: There is no guarantee that untrusted fab or assembly will perform IC testing correctly, or even at all. Such defective parts may exhibit correct functionality for the most part and therefore be very difficult to spot in

the supply chain. Untrusted foundry, assembly, and other third parties can use the rejected or defective components to artificially increase yield or sell them on the open market themselves. These components can pose a serious threat to the quality and reliability of any system that incorporates them [3–5, 8, 14].

11.3 Background

11.3.1 Related Work

There has been extensive research to combat theft, cloning, and counterfeiting of ICs committed by untrusted foundry. Active metering, logic obfuscation, source code encryption, and bitstream encryption for FPGAs are the main existing solutions to mitigate these attacks [9, 15–22]. Many of these schemes rely on "encryption" of combinational logic and/or finite state machine (FSM) blocks that function as locking mechanisms [15–17, 19–22]. In the case of locking mechanisms, only a specific input vector (i.e., a unique key) unlocks a new IC so that it functions correctly. To achieve this, extra logic blocks are inserted in the main design that only become transparent with a valid key. For example, a group of extra finite states are added in order to lock the FSM and only valid input sequence can bring the modified FSM to the correct initial state in normal working mode.

Active metering [15, 16, 20, 21] allows the IP owner to lock and unlock each IC remotely. The locking mechanism is often a function of the unique ID generated for each IC by a physically unclonable function (PUF, [23, 24]). Only the IP owner knows the transition table and can unlock the IC from this ID. In EPIC [21], each IC is locked with randomly inserted XOR gates. The XOR gates will only become transparent with the application of valid key (effectively unlocking the IC). In this technique, a set of public/private keys needs to be generated by the IP owner, foundry, and each IC. The primary objective of these approaches is to give the IP owner control over the exact number of ICs that can enter the market by obfuscating the correct behavior.

11.3.2 Challenges

The above techniques address only part of the IC counterfeiting problem, i.e., attacks by untrusted foundry, and completely ignore untrusted assembly. The active metering techniques described above attempt to prevent counterfeits from being produced; however, these techniques do not prevent the sourcing of out-of-spec/defective parts because they require that the IC be activated before it can be tested. The IP owner is required to provide the "key" to the IC before they know that the IC is not defective and is within specification. This can allow the untrusted foundry to ship/sell defective or out-of-spec ICs, which have already been activated by the IP owner. In addition, a foundry can request more keys than necessary from the IP owner by claiming a low yield. Thus, the foundry can still overproduce to some extent and place additional functional (defect-free) ICs in the market. Similar actions can be repeated by the assembly responsible for packaging, testing, and shipping the ICs to the market. Furthermore, there is no guarantee that untrusted fab or assembly will perform IC testing correctly, or even at all. Such defective parts may exhibit correct functionality for the most part and therefore be very difficult to spot in the supply chain. To summarize, the shortcomings in the prior approaches can potentially allow an untrusted foundry/assembly to ship cloned, overproduced, and out-of-spec/defective ICs to the market.

11.4 Connecticut Secure Split-Test

11.4.1 Overview

Typically, IP owners have little interaction with their ICs after providing GDSII files, test patterns, and corresponding responses to the foundry. After leaving the foundry, tested die are sent to the assembly to be packaged and tested again. Often, the ICs are sent directly from assembly to the market or to the IP owner's vendors. It is the responsibility of the assembly to ship only good ICs to the market in the volume requested by the IP owner. Provided this background, the primary goal of Connecticut Secure Split-Test (CSST) is to allow IP owners to reassert control over these processes without being physically present at the foundry or assembly.

CSST addresses the untrusted production flow in two dimensions, (i) adding CSST structures to the original design and (ii) providing communication from IP owner to foundry/assembly. Through the added structures, each IC is locked and its test responses are uniquely perturbed. The structures combined with the communication protocols allow only the IP owner to examine test responses and decide which ICs pass and which should be discarded. An IC will only become functional once the IP owner has decided that it has passed the necessary tests at assembly. Then and only then will the IP owner send a key back assembly to unlock the IC and make it useable. Through CSST, out-of-spec/defective ICs are prevented since the IP owner is the only entity with the authority and knowledge to pass an IC. If an IC is still locked, it will be nonfunctional and clearly visible if found in the supply chain. Overproduction can be prevented by how many keys are provided to the foundry/assembly (thus limiting the number of passing, unlocked ICs). Finally, cloning is prevented because only the IP owner can provide the correct key to unlock (use) the IC. Otherwise, the IC will remain nonfunctional.

Locking and unlocking of the ICs is provided by functional-locking and scan-locking blocks. The functional-locking block's purpose is to ensure that only unlocked ICs will have the correct functionality. The scan-locking block is used to

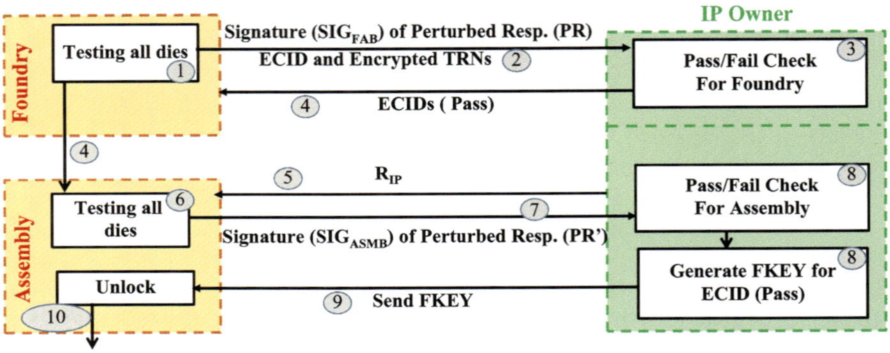

Fig. 11.2 Connecticut Secure Split-Test work-flow to prevent IC piracy

perturb the test response so that an outsider cannot determine the true test response from even an unlocked IC. The detailed steps of CSST are as follows (illustrated in Fig. 11.2)

1. At the foundry, each IC in a wafer generates a random number (TRN) and stores it internally in a one-time programmable (OTP) device. TRN is used by the functional-locking block to lock the functionality and the scan-locking block to internally perturb outputs from the scan chain. The TRN value is unique for each IC and results in different perturbed responses for different ICs. Each IC encrypts its TRN value internally using the IP owner's public key and outputs the encrypted TRN value to the fab. Hence, the foundry does not know the unique TRN value which prevents the foundry/assembly from guessing the true responses of the device.
2. The foundry applies test patterns to each wafer and collects the following information: electronic chip IDs (ECIDs) of all die, encrypted TRNs (ciphertext), and signatures from perturbed outputs of the scan chain. This information is sent to the IP owner.
3. With this information, the IP owner determines which die in the wafer pass/fail as follows. First, the TRN value is decrypted using the IP owner's private key. Note that since this key is only known to the IP owner, only the IP owner can determine the TRNs for each die. Next, the IP owner will compute the signatures associated with these TRNs. Finally, the IP owner will compare the signatures computed to those sent by the fab. Those die with signatures that match the IP owner's are considered fault-free.
4. The foundry marks passing die based on the IP owner feedback and sends the wafers to assembly where the wafers will be diced, packaged, and re-tested.
5. The IP owner sends a random number (R_{IP}) to the assembly that re-perturbs the test outputs. This step prevents the assembly from replaying the correct perturbed responses for passing ICs (obtained from fab or before packaging) and sending them to the IP owner without testing the IC.

6. The assembly applies the random number to the IC and receives the corresponding response after testing. The response is different than the response from the foundry because of the IP owner generated random number.
7. The assembly sends the signature from the perturbed responses with corresponding ECIDs to the IP owner for a decision. Data for all ICs are sent in a single session.
8. The IP owner checks the signature for each IC and decides which chips are functionally correct. The IP owner also generates keys to unlock the passing ICs. Note that these keys are a function of the TRN generated in Step 1 and the random number generated in step 5 and are therefore unique to each IC.
9. A single message containing the ECIDs of the passing die and their associated keys are sent to the assembly. Note that the IP owner can limit the number of keys sent out thus preventing overproduction.
10. The assembly/distributor applies the keys (FKEYs) sent by the IP owner (stores it in another OTP) to unlock functional block of the passing ICs. The key for each IC is different. The ICs that are still locked are useless to the assembly and clearly non-functional, thereby making them easy to detect if inserted into the supply chain.

11.4.2 CSST Structure

CSST is composed of both functional-locking and scan locking blocks. The functional locking mechanism ensures that the correct functionality of the IC is not revealed in order to prevent IC piracy by ensuring that only unlocked ICs will have the correct functionality. The scan-locking block ensures an untrusted party cannot scan out test results from an IC in an attempt to modify, bypass, or attack the CSST hardware. Even when the IC has received its functional key and is functioning correctly, the scan-locking block will prevent any attacker from applying patterns and observing the IC's responses.

11.4.2.1 Functional-Locking Block

The functional-locking block is used to lock the functionality of an IC to prevent IC piracy and out-of-spec/defective counterfeits. Basically, an IC that is locked will operate very differently than it's supposed to, thereby making it easy to spot in the supply chain (if defective) and impossible to use (if cloned). Figure 11.3 shows the functional-locking block of CSST. The functional-locking block consists of XOR_F mask, which is a series of m 3-input randomly inserted XORs, a TRNG, and two OTPs (OTP1 and OTP2).

XOR Mask The "XOR_F mask" is a series of m 3-input XOR gates which are inserted into non-critical paths of an, otherwise, unmodified circuit. XOR_Fs have

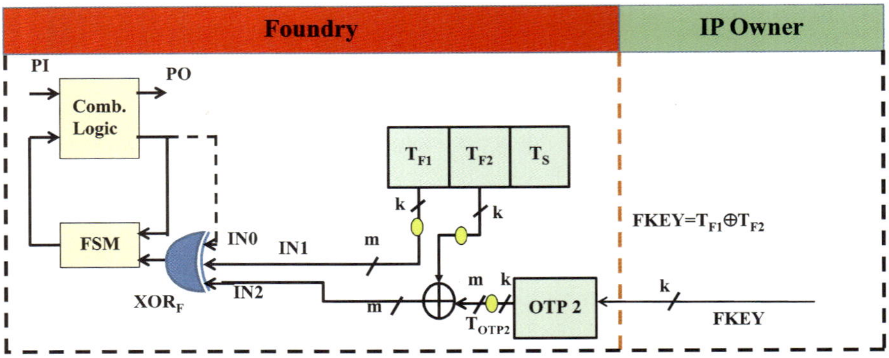

Fig. 11.3 Functional-locking block to lock the functionality of an original IC

three inputs IN0, IN1, and IN2. While IN0 is connected to circuit paths, XOR_Fs receive m-bit inputs as IN1 and IN2 with potential to modify the circuit. If the two inputs, IN1 and IN2, are the same, that particular part of the XOR_F mask will act as a buffer. If the two inputs are different, the XOR_F will act as an inverter. The placement of these XOR_Fs into the circuit dictates how they will affect the circuit. Hence, XOR_Fs are placed at the inputs of the scan flip-flops in the circuit. Random XOR_Fs at the scan flip-flop inputs, as shown in Fig. 11.3, invert the circuit's response as it is being captured by the scan flip-flops. The effect of having the inverting XOR_Fs at the scan flip-flop inputs is that some scan flip-flops may be capturing an inverted value of the actual response. Exactly which flip-flops are affected is determined by the two m-bit inputs IN1 and IN2. This property is useful as it means that the ICs can still be tested, and the test results are related to IN1 and IN2. Knowing the values for IN1 and IN2 allows the IP owner to know which test outputs should be inverted. Finding the correct test responses requires simple bit flipping and has negligible test time overhead.

True Random Number Generator Various true random number generators (TRNGs) have been designed [25, 26] for insertion in ICs. In digital circuits, TRNGs use physical phenomena such as clock jitter, temperature, power supply noise, etc., as a source of entropy. True Random Numbers (TRNs) cannot be predicted or algorithmically created by an attacker, even if the attacker has access to the design. An important quality of a TRNG is its unpredictable randomness. Since TRNGs have no stability requirement, they are usually smaller, and less complex than PUFs. TRN will be different for each IC but must remain constant throughout the IC's lifetime. However, TRNGs output different TRNs every time they are accessed and PUFs will not provide a stable and unique output every time it is activated due to its sensitivity to noise, temperature, and aging [23, 24]. To address this issue, the first time TRNG is accessed after manufacturing the TRN value will be stored into a one-time programmable memory (OTP) or polyfuse; the XOR_F input IN1 will therefore be connected to this memory rather than TRNG directly. Using OTP to store TRN solves the issue of stability with TRNG or PUF is nonexistent since the value in memory will always be constant. With an m-bit XOR_F, mask IN1 must be an m-bit random value. In order to reduce memory size and area overhead, a smaller k-bit random value can be used, k-bit value is then be repeated p times to enable m XOR_Fs ($m = p.k$).

RSA Asymmetric Encryption The RSA asymmetric cryptographic algorithm is a public-key cryptographic system, which means that the encryption and decryption processes are performed using different keys. Hence, we use an industry-verified secure implementation of the RSA algorithm that implements the many security standards relating to the algorithm, such as the PKCS #1 standard [27]. During manufacturing, an RSA public key is embedded into the design in read-only memory; this public key will be the same for all circuits. Due to the asymmetric nature of RSA, using the same public key or allowing the foundry to read the public key does not pose a risk to CSST as [28] and [29] have proven that computing the RSA private key from its corresponding RSA public key is as complex as factoring RSA's modulus n into its prime factors.

Working Principle TRN is stored in OTP1 so that it remains constant over the IC's lifetime. TRN is divided into three parts, T_{F1}, T_{F2}, and T_S. The T_{F1} and T_{F2} are used to lock the IC functionality via XOR_F. An XOR acts like a buffer when it's other input experiences '0' and acts like an inverter with '1'. T_S is used to control a scrambling block in the modified scan-locking block (see Sect. 11.4.2.2). A 3-input XOR_F is used to send the value of circuit path directly or inversely for functional locking. IN0 and IN1 are connected directly to the circuit path and T_{F1} respectively. OTP2 is initially set to all 0s or all 1s which is known to the IP owner. The T_{F2} and output of OTP2 are XORed and connected to IN2. Initially, the value of OTP2 is all 0s or all 1s. The contents of OTP2 are XORed with T_{F2}. Hence, IN2 receives either T_{F2} or T'_{F2} depending on all 0s or all 1s in OTP2. Depending on T_{F1} and T_{F2}, the $XORF$s act like inverters or buffers. IN1 and IN2 are made different so the IC remains locked. To unlock the IC, IP owner sends an $FKEY$ in such a way that IN1 and IN2 receive same value and $XORF$s become transparent. Only $T_{OTP2} = FKEY = T_{F1} \oplus T_{F2}$ satisfies this condition. $FKEY$ can be generated only by the IP owner who knows TRN, is unique for each chip, and it does not reveal any information about TRN. Note that in this design, one can trade-off OTP size with security by (1) using smaller k-length T_{F1}, T_{F2}, and (2) OTP2 broadcasting m XOR_Fs p times such that $m = p * k$ where p is an integer.

11.4.2.2 Scan-Locking Block in CSST

The scan-locking block is used to perturb the test response so that an outsider cannot determine the true test response from even an unlocked IC. Figure 11.4b shows the scan-locking block. The yellow blocks represent the scan-chain of the design and the test structure commonly used in practice while the rest are required to implement CSST. Test data compression is required to overcome

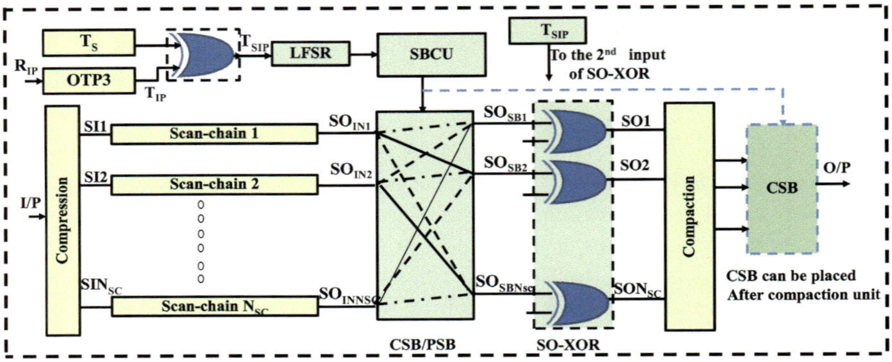

Fig. 11.4 Scan-locking block to enhance security

the limitations of the automatic test equipment (ATE). The outputs of the scan chains are scrambled through a scrambling block (SB) in order to perturb the functionally locked/unlocked response. The output of scrambling block (SB) is sent through SO-XOR blocks for further perturbation. Scrambling blocks are an essential component in telecommunication and microprocessor communication [30].

The scrambling block can be completely shuffled or partially shuffled. Complete scrambling block (CSB) is designed such that all inputs to the block can potentially go to any output pin received by a compaction circuit. On the other hand, a partial scrambling block (PSB) is designed in such a way that an input to the scrambling block is connected with only N_{SB} different outputs. Non-blocking crossbar switch [30] is a strong candidate for the scrambling block and can be designed with pass transistors or transmission gates. The security strength depends on the type of shuffling block. A CSB will provide maximum security but higher cost. Alternatively, by using PSB, N_{SB} can be tuned with the consideration of area overhead and desired security strength.

The scrambling block's controlling unit (SBCU) assures that all inputs to the scrambling block, either CSB or PSB, are seen at the output. The logic of SBCU depends on the PSB or CSB structure and number of scan chains (N_{SC}). The output of LFSR, which controls the SBCU, changes in each clock cycle and depends on initial seed, T_{SIP}=T_S $\oplus$ T_{IP}, which is known by the IP owner only. T_{IP} is the value stored in OTP3. In foundry, T_{IP} is set to all 0s or all 1s. But in assembly, the IP owner sends a random number, R_{IP}, independent to IC and TRN value. The initial seed of LFSR, T_{SIP}, is different for assembly and foundry for the same IC; hence SBCU performs different scrambling.

The output of CSB/PSB is sent through an SO-XOR block to add another layer of security. The scrambling block has to be ready before intermediate output of scan chain, SO_{IN}, is ready. In order to avoid timing failure, LFSR can be activated in the negative clock edge so that scrambling block is ready and SO_{IN} can pass through it. The SO-XOR block is controlled by T_{SIP}. Depending on the second input of

XOR in SO-XOR block, the output of scrambling block flips or goes transparent. The scrambling block and SO-XOR block make it impossible for untrusted fab and untrusted assembly to determine the correct output responses.

Alternative Placement of SB The size of scan chains increases with gate-counts and flip-flops but ATE has only limited number of channels. Compaction circuitry, such as MISR, is commonly used to compact the scan-chain responses and support the ATE. The CSB in CSST provides the best security level but requires large area. The cost reduces with the size of CSB; for example, a 10×10 complete shuffle crossbar switch requires 100 transmission gates whereas a 4×4 requires 16. In order to reduce the cost with lesser impact on security, CSB can be placed after the compactor (we refer to this as alternative place). An $r : 1$ compactor reduces r^2 times area for scrambling block.

11.4.3 Experimental Results and Analysis of CSST

In this section, we discuss the security provided by CSST and CSST's resistance to attacks.

11.4.3.1 Security Enhancement

CSST has important security features such as tunable locking blocks. CSST uses two layers of security for both locked and unlocked ICs. The first layer consists of the scrambling block that permutes the output of the scan chains in a different way for every test pattern based on the LFSR seed, T_{SIP}. The second layer contains XORs between MISR and output of original scan chains. The first and second layers of CSST can be tuned to trade off security and cost. The scrambling block could be a CSB for highest security but with highest cost. The scrambling block could also be a PSB with variable N_{SB}. Smaller N_{SB} reduces the number of possible permutations which decreases the security. Designing SB to minimize cost will be investigated in future work. The number of XORs in the second layer can be tuned in both CSST and CSST. Below we perform experiments to capture the security and tradeoffs between these parameters.

The hamming distance (HD) is a popular metric to analyze the security strength. The average %HD between the actual response and captured response from modified scan-locking block is presented in Tables 11.1 and 11.3. A $\sim 50\,\%$ HD is hard to predict and represents high security while $\sim 0\,\%$ or $\sim 100\,\%$ HD is easy to predict. The IC was a synthesized implementation of the ISCAS'89 benchmark s38417 and ITC'99 benchmark b19. The result shows that the modified-scan-locking block can perturb the actual responses regardless of the size of the IC. s38417 and b19 have 10 and 50 scan chains respectively. Security is enhanced in CSST by both shuffling block and SO-XOR block.

Table 11.1 Hamming distance comparison for different number of XORs in SO-XOR block ($N_{SB} = N_{SC}/2$ i.e. 50 % of total scan chains are passed through scrambling block)

N_{SO-XOR} (10–50 % of N_{SC})		%HD					
		CSST		SB Before Compaction $N_{SB} = N_{SC}/2$		SB after Compaction	
s38417	b19	s38417	b19	s38417	b19	s38417	b19
1	5	9.06	9.91	29.29	38.18	42.87	43.76
2	10	19.66	13.61	40.01	43.69	46.41	48.12
3	15	22.89	20.54	48.73	49.31	47.12	47.78
4	20	25.79	26.66	47.44	49.31	49.31	48.89
5	25	36.36	36.43	50.03	49.31	50.03	50.00
6	30	46.46	39.87	45.63	50.00	49.31	48.89
7	35	47.44	43.69	47.44	49.31	50.03	50.00
8	40	49.31	48.89	50.03	50.00	50.03	50.00

Table 11.2 %HD analysis of CSST with CSB at alternative place for different compaction ration, r, of a compactor $N_{SO-XOR} = N_{SC}/2$

Benchmark	s38417		b19	
Compaction ratio, r	2	5	5	10
%HD (Foundry)	42.04	44.59	48.41	49.04
%HD (Foundry, Assembly)	28.17	36.47	39.82	42.04

Table 11.3 Hamming distance analysis for different scrambling blocks in CSST with $N_{SO-XOR} = N_{SC}/2$

Benchmark	s38417				b19			
N_{SB}	2	3	5	10	5	10	25	50
%HD (Foundry)	42.04	44.59	50.03	50.03	48.07	48.84	49.31	50.00
%HD (Foundry, Assembly)	28.17	36.47	39.82	42.04	31.84	37.81	44.59	48.41

Table 11.1 shows the effectiveness of scrambling block in CSST. The result shows that for $N_{SB} = N_{SC}/2$ there is a huge shift of hamming distance to ideal value (50 %). The location of SB impacts the security of whole system. The SB can be placed between the scan chain and compaction circuit or after the compaction unit (alternative place) to reduce the area overhead. The result shows that SB in alternative place gives better hamming distance due to the effectiveness of shuffling block.

Table 11.2 shows the effect of placing SB at alternative place for different compact ratio of MISR (compactor). Placing SB at alternative place offers high quality security with lower area overhead. The total number of XORs, N_{SO-XOR}, in SO-XOR block were varied to understand the effectiveness of SO-XOR block. The result shows that different N_{SO-XOR} values are required for different benchmarks.

Table 11.3 shows the effectiveness of scrambling block and R_{IP}. The results show that $N_{SB} = 10\,\%\ N_{SC}$ can achieve ideal hamming distance. Same die gives different responses from foundry to assembly because the random number,

R_{IP}, provided by IP owner scrambles the response differently. The last row of Table 11.3 shows that foundry and assembly possess significant hamming distance between responses for same IC. The scrambling block usually takes large area in CSST but high security can be achieved by tuning SO-XOR block and scrambling block. Tables 11.1 and 11.3 show that the desired security with lowest area can be achieved by activating both PSB and partial SO-XOR block. The result shows that $N_{SB} = 50\,\%\ N_{SC}$ and $N_{SO-XOR} = 50\,\%$ of other N_{SC} ensures maximum security.

11.4.3.2 Attack Analysis

Overall, CSST significantly increases the security of the IC supply chain. It is worth analyzing the possible attacks on this technique and the security that CSST provides. Different attacks would include: (1) attacks on the design (direct attacks), (2) attacks to modify the netlist (tampering attacks), (3) attacks which attempt to deceive the IP owner or avoid the technique (circumvention attacks), (4) hardware-based attacks that attempt to remove the CSST blocks (removal attacks), and (5) Unlocked IC attacks.

Direct Attacks Connecticut Secure Split-Test is relatively resilient to direct attacks. Each IC requires one key to reach a fully functional state, and from a hardware perspective it would be easy to have a single output pin which indicates whether or not the IC is in that state. The problem of finding a key that puts the IC into a fully functional state is equivalent to the problem of bypassing RSA. An attacker who tries to bypass this technique would have two options: (i) randomly generate potential keys in the hopes that they find one which works for a known TRN, or (ii) factor the public modulus into its component primes so that they can find the private key themselves and instantly generate the correct key. Both of these attacks are proven to be difficult. The first one would require billions of iterations since a key of length $x$would have 2^x possibilities. The second attack is equivalent to attacking RSA which has not been successful yet and is extremely difficult as shown in [31].

Circumvention Attacks Attacks which try to bypass CSST do not fully defeat the technique. If an attacker has knowledge of which XOR_Fs and SO-XORs in the circuit have been activated by the key/TRN combination used, the attacker could figure out which responses have been inverted. The attacker could change the responses obtained in order to make the IC pass the test. The IP owner would then pass the. This attack could be done at three different stages: (i) at the foundry, (ii) at the assembly, or (iii) both foundry and assembly working together. If the attack occurs at the foundry, the bad IC will be sent to the assembly, where the same tests are applied using the same TKEY. This time, the IC will fail and the IP owner knows to discard the IC. Additionally, the IP owner will see that the foundry sent a failing IC to the assembly, raising a red flag about the integrity of the foundry. If the attack occurs at the assembly, the IP owner will be able to know results are being changed because the same tests were done by the foundry using the same TKEY.

Any mismatch in results between foundry and assembly can detect an attack on the IC and the IP owner will know to discard and not send a final FKEY for that IC. In the third case of collusion between foundry and assembly, using the same TKEY will not prevent the attack. However, if different TKEYs are used for the foundry and assembly, the attacker has to figure out the difference between the TKEYs in order to know which outputs to change. This task is the same as cracking RSA which, as mentioned before, is infeasible for large keys.

Tampering Attacks An attacker could try to re-route the outputs from TRNG to go directly to the output of RSA, bypassing the RSA decryption needed to activate the IC. We acknowledge that this attack could compromise the system; however, rather than defining separate blocks for each component, the components can be synthesized and optimized with the design in order to obfuscate the components and their functionality, e.g., generate a flat netlist. The XOR_F mask can also be hidden into the design during synthesis. To prevent the attacker from figuring out the location of the XOR_F mask, other gates or a combination of gates (NAND, NOR, etc.) can be used instead of XORs. Using other gates would not change the design or effectiveness of CSST, but would make it almost impossible to find which gates belong to CSST. Obfuscation makes the task of finding the individual CSST components difficult and helps prevent attacks. However, it is important to note that attacks to re-route nets or bypass gates from the XOR_F mask require a high degree of sophistication such as high technical background, access to the design files, knowledge of circuits functionality, multiple resources, and time. These aspects of the attack alone deter many of the counterfeit attacks because current overproduction or selling of defective ICs does not require any level of sophistication.

Removal Attacks It is possible that the foundry may try to remove some or all of the hardware needed by this technique. Exactly how much they can remove depends on how much they know about the logic design of the IC. For example, they could not blindly remove any XOR gate whose output connects to a flip-flop input; they would have to know whether or not the XOR gate was part of the XOR_F mask. Attacks aimed to tamper with or remove the TRNG block or RSA block would have to be very carefully designed to avoid detection. This is especially true because, as specified, this technique implements a basic metering methodology that requires foundries to report each IC back to the IP owner and requires that the IP owner provide a working key for the IC. Attacks that altered the way that the TRNG or RSA blocks worked would also have to avoid communications with the IP owner, as the TRNG and RSA blocks directly affect the scan output during testing.

Unlocked-IC Attacks CSST is resilient to attacks using unlocked ICs. With the scan-locking block, the IC manufacturers cannot run the test vectors on the unlocked IC to determine the response as the inputs of some scan chains are inverted using SCB. As a result, an attacker cannot obtain any information from an unlocked IC.

Flush and Shifting Attack Flush test is performed to find any defects in a scan chain. In flush test, a selected flush pattern (e.g. 11001100 or 111111, or 000000) is shifted all the way through scan chain input and expect to get the same pattern arriving at the output. An attacker might try to shift in all 0s or all 1s to obtain the functionality of scan-locking block (specifically, T_S), but CSB and PSB make it nearly impossible.

Graph Isomorphism Attack Liu and Wang [9] described that this potential attack can reveal the locking mechanism. But in CSST, an adversary does not know the functionality of the logic network as he/she does not get the exact test response from testing because of scan-locking block. Another potential attack is that the untrusted foundry will send random encrypted TRN to get idea about correct response. However, whenever IP owner receives the wrong TRN, it won't pass the IC. Several attempts could be viewed as low yield and hence harm the reputation of the foundry/assembly.

OTP Attack An adversary might change the OTP value to all 0s or to all 1s to make the XOR_F transparent. To prevent this attack, XOR_F block is implemented by inserting both XOR and XNOR randomly. In order to prevent such attacks in scan chain, LFSR can also be designed in such a way that for all 0s it does not output all 0s and SO-XOR block might be built using both XOR and XNOR.

11.4.3.3 Overhead and Coverage

Coverage Analysis Due to the added circuitry, there will be additional faults introduced to the IC. Fault simulation done on RSA has shown that this circuit can achieve 95.83 % test coverage with only 960 random patterns. These results show that the same random tests used on the IC in built-in self-test (BIST) mode can be used to test RSA and obtain a high fault coverage. Note that generally it is recommended to use BIST for testing cryptographic hardware in a secure IC because of its increased security over test application from tester [32]. As for TRNG, no test is necessary to detect faults in it. The purpose of TRNG is to create random values, these values are stored in memory. Any faults in TRNG do not matter since its output is random. The XOR_F masks will introduce four faults per XOR after fault collapsing. Since the XOR_F mask is inserted at the input of scan flip-flops, i.e., at the output of the combinational logic, these faults are highly observable , which indicates they are closest to the scan flip-flops capturing their responses and are easy to detect.

Test Time Overhead Analysis CSST does not need much extra test time because of simple communication between IP owner and foundry/assembly. The entire process easily lends itself to automation as well.

Area Overhead We calculated the area overhead as follows. RSA can take k-bit input and gives k-bit output provided that the length of the public key, K_{pub}, is also k bits. RSA encryption's throughput and speed is higher than RSA decryption and requires less area overhead [28]. The modified functional-locking block requires m XOR_Fs where $m = p * k$ and p is an integer. Each of k-bit T_{F1} and k-bit T_{F2}, and k-bit T_{OTP2} can be broadcast p times in order to connect all m XOR_Fs. Ring oscillator based TRNG is easy to implement and does not need to add additional circuitry as k increases. The size of LFSR depends on the size of scrambling block and the size of scrambling block depends on N_{SB} and total number of scan chains, N_{SC}. $N_{SB} = N_{SC}$ for a CSB and $N_{SB} < N_{SC}$ for a PSB. The size of SBCU depends on N_{SC} and structure of scrambling block.

Table 11.4 shows the area overhead of CSST. The result shows that CSST is a strong function of N_{SB} and N_{SC}. The size of scrambling block, LFSR, and SBCU depend on the number of scan chains, N_{SC}. The main cost of modified scan-locking block is scrambling block and SBCU. The area overhead can be reduced by placing CSB after compaction unit. The area overhead for placing CSB at alternative place depends on the compact ratio of a compactor. The result shows that the area overhead is reduced significantly if CSB is placed after compactor.

11.5 Summary

The emergence of a horizontal business model for IC production has led to new vulnerabilities in the supply chain. Once a design gets to the foundry/assembly stage, the design house has minimal, if any, control over what the foundry/assembly does or does not do. Hence, if left unchecked, untrusted foundries and assemblies can potentially source overproduced, defective/out-of-spec, and cloned ICs. In order to combat IC piracy and issues of trust between the design house and the foundry/assembly, countermeasures such as hardware metering were introduced. Such techniques require that the ICs produced in the foundry/assembly be "unlocked" by the design house. Unfortunately, such techniques unlock the chips before they are tested by the foundry and assembly. Hence, the design house would not be able to truly control the number of (nondefective or defective) ICs that enter the supply chain. The CSST technique, however, combats all these IP piracy issues by allowing the design house to both authenticate and verify the IC functionality. By incorporating structures such as functional locking blocks and scan-locking blocks into the design of the IC and enabling an effective communication flow, the design house can "see" the ICs as they are being tested in the assembly/foundry and decide whether to accept or reject an IC. No one but the IP owner can differentiate between passing and failing dies, and unlock passing chips. Those chips that do not pass testing and remain locked are clearly visible in the supply chain. In addition, CSST offers better security in terms of resistance to attacks/circumvention and more tunable parameters to scale the security than prior approaches.

Table 11.4 Area overhead for a 10M gates design with $k = 1024$ and $N_{SC} = 1000$

	CSST	SB before Compaction			SB after Compaction	
m	(%)	$N_{SB} = 10$ (%)	$N_{SB} = 100$ (%)	$N_{SB} = 1000$ (%)	$r = 50$ (%)	$r = 100$ (%)
1024	0.0222	0.0366	0.1365	1.1355	0.0244	0.0366
2048	0.0233	0.0377	0.1376	1.0366	0.0255	0.0377
5192	0.0266	0.04	0.1399	1.1389	0.0289	0.04
10240	0.0322	0.0466	0.1576	1.5762	0.0344	0.0466

References

1. M. Tehranipoor, H. Salmani, X. Zhang, *Integrated Circuit Authentication: Hardware Trojans and Counterfeit Detection* (Springer, New York, 2014)
2. U. Guin, D. DiMase, M. Tehranipoor, Counterfeit integrated circuits: Detection, avoidance, and the challenges ahead. J. Electron. Test. **30**(1), 9–23 (2014)
3. U. Guin, D. Forte, M. Tehranipoor, Anti-counterfeit techniques: from design to resign, in *Microprocessor Test and Verification (MTV)*, 2013
4. M. Rahman, D. Forte, Q. Shi, G. Contreras, M. Tehranipoor, Csst: An efficient secure split-test for preventing ic piracy, in *North Atlantic Test Workshop (NATW), 2014 IEEE 23rd*, May 2014, pp. 43–47
5. G. Contreras, M.T. Rahman, M. Tehranipoor, Secure split-test for preventing IC piracy by untrusted foundry and assembly, in *Proc. International Symposium on Fault and Defect Tolerance in VLSI Systems*, 2013
6. M. Rostami, F. Koushanfar, R. Karri, A primer on hardware security: Models, methods, and metrics. Proc. IEEE **102**(8), 1283–1295 (2014)
7. U. Guin, K. Huang, D. DiMase, J. Carulli, M. Tehranipoor, Y. Makris, Counterfeit integrated circuits: A rising threat in the global semiconductor supply chain. Proc. IEEE **102**(8), 1207–1228 (2014)
8. M. Rahman, D. Forte, Q. Shi, G. Contreras, M. Tehranipoor, CSST: preventing distribution of unlicensed and rejected ICs by untrusted foundry and assembly, in *Proc. International Symposium on Fault and Defect Tolerance in VLSI Systems*, 2014
9. B. Liu, B. Wang, Embedded reconfigurable logic for asic design obfuscation against supply chain attacks, in *Design, Automation and Test in Europe Conference and Exhibition (DATE), 2014*, March 2014, pp. 1–6
10. A. Yeh, Trends in the global IC design service market, (2012) http://www.digitimes.com/news/a20120313RS400.html?chid=2
11. R. Maes, D. Schellekens, P. Tuyls, I. Verbauwhede, Analysis and design of active ic metering schemes, in *IEEE International Workshop on Hardware-Oriented Security and Trust, 2009. HOST '09*, 2009, pp. 74–81
12. R. Torrance, D. James, The state-of-the-art in ic reverse engineering, in *Proceedings of the 11th International Workshop on Cryptographic Hardware and Embedded Systems*, ser. CHES '09 (Springer, Berlin, Heidelberg, 2009), pp. 363–381. [Online]. Available: http://dx.doi.org/10.1007/978-3-642-04138-9_26
13. I. McLoughlin, Secure embedded systems: The threat of reverse engineering, in *14th IEEE International Conference on Parallel and Distributed Systems, 2008. ICPADS '08*, Dec 2008, pp. 729–736
14. U. Guin, M. Tehranipoor, D. DiMase, M. Megrdichian, Counterfeit IC detection and challenges ahead. ACM/SIGDA E-Newsletter **43**(3), (2013)
15. Y. Alkabani, F. Koushanfar, M. Potkonjak, Remote activation of ICs for piracy prevention and digital right management, in *Proc. of IEEE/ACM International Conference on Computer-Aided Design*, 2007, pp. 674–677
16. F. Koushanfar, Provably secure active ic metering techniques for piracy avoidance and digital rights management. IEEE Trans. Inform. Forensics Secur. **7**(1), 51–63 (2012)
17. R. Chakraborty, S. Bhunia, HARPOON: An obfuscation-based SoC design methodology for hardware protection. IEEE Trans. Comput. Aided Des. Integrated Circ. Syst. **28**(10), 1493–1502 (2009)
18. R. Chakraborty, S. Bhunia, Hardware protection and authentication through netlist level obfuscation, in *Proc. of IEEE/ACM International Conference on Computer-Aided Design*, November 2008, pp. 674–677
19. R. Chakraborty, S. Bhunia, Rtl hardware ip protection using key-based control and data flow obfuscation, in *VLSID '10. 23rd International Conference on VLSI Design, 2010*, Jan 2010, pp. 405–410

20. J. Rajendran, M. Sam, O. Sinanoglu, R. Karri, Security analysis of integrated circuit camouflaging, in *Proceedings of the 2013 ACM SIGSAC Conference on Computer & Communications Security*, ser. CCS '13 (ACM, New York, NY, USA, 2013), pp. 709–720. [Online]. Available: http://doi.acm.org/10.1145/2508859.2516656
21. J. Roy, F. Koushanfar, I. Markov, EPIC: Ending piracy of integrated circuits, in *Proc. on Design, Automation and Test in Europe*, March 2008, pp. 1069–1074
22. A. Baumgarten, A. Tyagi, J. Zambreno, Preventing IC piracy using reconfigurable logic barriers. IEEE Design Test Comput. **27**(1), 66–75 (2010)
23. G. Suh, S. Devadas, Physical unclonable functions for device authentication and secret key generation, in *44th ACM/IEEE Design Automation Conference, 2007. DAC 07*, June 2007, pp. 9–14
24. M. Rahman, D. Forte, J. Fahrny, M. Tehranipoor, Aro-puf: An aging-resistant ring oscillator puf design, in *Design, Automation and Test in Europe Conference and Exhibition (DATE), 2014*, March 2014, pp. 1–6
25. M.T. Rahman, K. Xiao, D. Forte, X. Zhang, J. Shi, M. Tehranipoor, Ti-trng: Technology independent true random number generator, in *Proceedings of the The 51st Annual Design Automation Conference on Design Automation Conference*, ser. DAC '14 (ACM, New York, NY, USA, 2014), pp. 179:1–179:6. [Online]. Available: http://doi.acm.org/10.1145/2593069.2593236
26. B. Sunar, W. Martin, D. Stinson, A provably secure true random number generator with built-in tolerance to active attacks. IEEE Trans. Comput. **56**(1), 109–119 (2007)
27. RSA Laboratories, PKCS 1 v2.1: RSA Cryptography Standard, 2002
28. C. Mclvor, M. McLoone, J. McCanny, Fast montgomery modular multiplication and rsa cryptographic processor architectures, in *Conference Record of the Thirty-Seventh Asilomar Conference on Signals, Systems and Computers, 2004*, vol. 1, Nov 2003, pp. 379–384
29. Z. Keija, X. Ke, W. Yang, M. Hao, A novel asic implementation of rsa algorithm, in *Proceedings. 5th International Conference on ASIC, 2003*, vol. 2, Oct 2003, pp. 1300–1303
30. Y. Tamir, H.-C. Chi, Symmetric crossbar arbiters for vlsi communication switches. IEEE Trans. Parallel Distr. Syst. **4**(1), 13–27 (1993)
31. G.L. Miller, Riemann's hypothesis and tests for primality, in *Proceedings of Seventh Annual ACM Symposium on Theory of Computing*, ser. STOC '75 (ACM, New York, NY, USA, 1975), pp. 234–239. [Online]. Available: http://doi.acm.org/10.1145/800116.803773
32. D. Karaklajić, M. Knežević, I. Verbauwhede, Low cost built in self test for public key crypto cores, in *2010 Workshop on Fault Diagnosis and Tolerance in Cryptography (FDTC)*, Aug 2010, pp. 97–103

Chapter 12
Chip ID

To prevent the widespread infiltration of counterfeit parts, traceability of electronic components in the supply chain demands more attention. Due to globalization, these components are now manufactured and assembled across the world. Thus, it is necessary to trace the origin of a component to validate the authenticity of its manufacturer. It is clearly mentioned in SAE aerospace standard AS5553 [1] that users should require their suppliers to trace components back to the original component manufacturers (OCMs). However, traceability has been obscured due to (i) high complexity of the component supply chain. For example, there are thousands of different types of components with different types (analog, digital, and mixed-signal) and sizes (small, medium, and large) present in the supply chain [2, 3]; (ii) different cultures and national interests across the globe. The work cultures differ across countries. Thus, different rules and regulations are imposed while manufacturing components; and (iii) the lack of low-cost secure solutions to track and trace each component. The pre-existing markings on the components can be easily copied and reprinted on the component.

Traceability requires a unique identification number (ID) to track and trace each component throughout the supply chain. The ID can be marked on the die (die ID) or on the package (package ID) of a component. In this chapter, we will first discuss die ID using physically unclonable functions (PUFs) and their challenges and limitations. We will then introduce four different package IDs for the traceability of electronic components: encrypted QR codes, DNA markings, nanorods, and coating physical unclonable functions (PUFs). For each ID, we describe their challenges and limitations for the detection of counterfeit components.

M. Tehranipoor et al., *Counterfeit Integrated Circuits*,
DOI 10.1007/978-3-319-11824-6_12

12.1 General Requirements of Chip ID

To ensure the authenticity of a component, it is necessary to create an industry standard marking protocol for the chip ID. The chip ID should consist of two parts—fixed part and variable part. The fixed part should contain all the relevant information to identify the part type and its source (e.g., date/lot code, manufacturer ID, country of origin, etc. mentioned in Section 3.9.5 of MIL-PRF-38534H [4]). The variable part should contain a unique identification number to differentiate between two components of same and/or different types. To efficiently detect counterfeit components, the chip ID should satisfy the following criteria:

i. Uniqueness: This is a measure of uncorrelatedness or dissimilarity between two chip IDs. Ideally, the bits in two IDs should differ with a probability of 0.5 under the same test conditions. Making the IDs more unique reduces the aliasing effect, i.e., having the same ID for two or more components, in the component supply chain. Uniqueness does not help to detect counterfeit components as it does not provide any protection against copying and printing. However, it does provide a unique identity of a component in the supply chain.
ii. Unclonability: This is a measure of the difficulty to reproduce the same ID. A chip ID is unclonable if it is practically impossible to generate the same ID after observing one. If the counterfeiter can reproduce an ID, it would make the ID vulnerable to counterfeiting. Multiple copies of one ID could be printed on different counterfeit components. As a result, the adoption of such vulnerable IDs will be extremely risky as test escapes and counterfeit components could find their way into the supply chain. The unclonable property of IDs provides them with the necessary resistance against cloning.
iii. Manufacturability: The creation of chip IDs needs to be a stand-alone process. It should not drastically intervene with any manufacturing (fabrication and packaging) steps of a component. The generation of IDs should be seamlessly integrated with the manufacturing process. For package IDs, one should be able to evaluate the reliability of a component after the ID is printed on it.
iv. Reliability: The chip ID should be reliable and stable in all operating conditions, such as temperature, humidity, vibration, etc., specified for a component. Package IDs need to withstand the temperature variations specified for the device class. For example, the marking should remain intact during −55°C to 125°C for military grade components. The ID must be robust enough to travel with and remain unchanged during the entire length of component's lifetime.
v. Cost Effectiveness: The chip ID should be cost effective. It should not incur considerable extra costs. Otherwise, universal adoption will not be possible due to the sheer number of parts, ranging in value from a few cents to hundreds of dollars, available in the component supply chain.
vi. Ease-of-Use: It should be easy to measure and/or decode all the relevant information in an ID. For example, package IDs should be verifiable with a simple hand-held measurement instrument and at any point along the supply chain.

12.2 Die ID

Techniques to generate die IDs are based on extracting unique features and parameters from a circuit to help uniquely identify each chip or embedding a unique ID into the chip during or after fabrication and test. The conventional approach includes writing the unique ID into a non-programmable memory, such as One-Time-Programmable (OTP), ROM, etc., or using post-fabrication external programming techniques, such as laser fuses [5] or electrical fuses (eFuses) [6]. However, the IDs generated by this method are static and vulnerable to different attacks, such as cloning and tampering. To mitigate this problem, physically unclonable functions (PUFs) were proposed to generate non-static IDs that are resistant to cloning and tampering. In this section, we will briefly describe different PUFs used for generating unique die IDs.

12.2.1 Physically Unclonable Functions (PUFs)

Random physical features such as fingerprints, which are unique to each individual and difficult to remove/duplicate, have a long history of use in biometrics. PUFs are analogous to fingerprints and have gained popularity in recent years. In essence, PUFs serve as "silicon fingerprint" that can uniquely identify each die/chip. As the PUFs are designed within the die, we categorized the IDs generated by PUFs as die IDs.

Silicon PUFs were first proposed by researchers at MIT in [7] as a way to identify ICs. Due to variations occurring in the manufacturing process, each fabricated instance of a design in silicon has slightly different physical features and performance characteristics. A silicon PUF is a special circuit embedded in an IC that extracts the IC's random characteristics to generate a unique signature, or identifier [8–10]. Before we discuss the operation of basic PUF structures, it is necessary to note some terminologies that are associated with PUFs. Inputs and outputs of PUF circuits are typically referred to as *challenges* and *responses* respectively. An applied challenge and its measured response are referred to as a *challenge-response pair (CRP)*. We refer to all the PUF response bits as the PUF *signature*.

Silicon PUFs have properties that make them exceptional candidates to thwart counterfeiting attacks [10]. First, since many of the fabrication variations in a die are random, the unique signature generated by a PUF cannot be cloned or replicated, even by the manufacturer. Thus, in order to obtain the PUF's signature, one must have or have previously been in physical possession of the IC containing the PUF. Second, the PUF technology is tamper resistant because any attempt to physically tamper with the IC may harm the IC's physical features and modify its associated performance characteristics. For example, if an attacker attempted to steal the PUF key through microprobing, the de-metalization and delayering steps would destroy or modify the key, thereby leaving the attacker empty-handed.

12.2.2 PUF Structures

There are two main types of silicon PUFs discussed in the literature [10]: delay-based PUFs and memory-based PUFs. Delay-based PUFs use race conditions to extract variations of wire and gate delays to generate PUF signatures. Examples include the arbiter PUF [8], Ring Oscillator (RO) PUF [8], and aging-resistant RO-PUF [11]. Memory-based PUFs exploit the random settling behavior of volatile memory elements to generate PUF signatures. SRAM PUF [9] is an example memory-based PUFs. In the subsections below, we will describe arbiter PUF, RO PUF, and SRAM PUF in detail.

12.2.2.1 Arbiter PUF

The Arbiter PUF [7] was the first silicon PUF realized in an IC. The arbiter PUF sets up two paths (designed symmetrically for same intended path delay) and uses a race condition to generate a 1-bit output (response) as follows. The two paths are simultaneously asserted with an input pulse. At the end of the paths, an "Arbiter" determines which asserted path won the race. If the pulse reaches the output of the first path faster, the Arbiter outputs a logic 1 (HI). Otherwise, it outputs a logic 0 (LO). The output/response depends on the delay present in both paths and is a function of the variations experienced by an IC during fabrication.

The Arbiter PUF structure is shown in Fig. 12.1b. Each path consists of a set of stages with each stage containing a switch circuit. The switch circuit is composed of two MUXs (see Fig. 12.1a) which are controlled by a challenge bit. The challenge bit determines which paths the input signals take within each switch. For example, with a challenge bit set as LO, the input signals will continue to the output along their current paths. When the challenge bit is set HI, the signals will switch paths. To illustrate, the paths to a particular challenge are shown in Fig. 12.1c. Due to the variations occurring in the manufacturing process, the delays of each path within the switches will vary among ICs. Hence, the propagation time through both of the selected paths is random. The arbiter at the end of the paths is typically implemented with a D-latch.

While the Arbiter PUF was the first PUF proposed in the literature, a robust Arbiter PUF is tough to achieve in practice. First, to generate a correct response, the timing difference between the two paths has to satisfy the setup time and hold time requirements of the D-latch. Second, the routing of both paths must be perfectly symmetric which can be difficult to obtain in practice [12], especially in FPGAs. Without symmetric routing, the PUF response bits are biased towards one value (LO or HI). Finally, it is been shown that after observing a number of CRPs, simple machine-learning techniques can be used to predict PUF responses to unseen challenges with relatively high accuracy [10]. This flaw could allow attackers to determine a PUF response to a new challenge without being in possession of the IC.

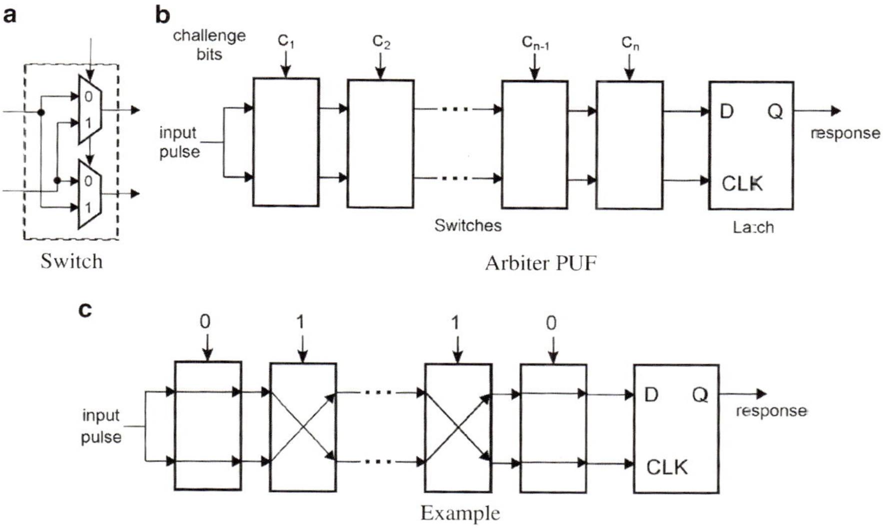

Fig. 12.1 (**a**) A switch block constructed with multiplexers (MUXs) controlled by a challenge bit; (**b**) The architecture of an arbiter PUF; (**c**) Effects of challenge bits on paths to the arbiter

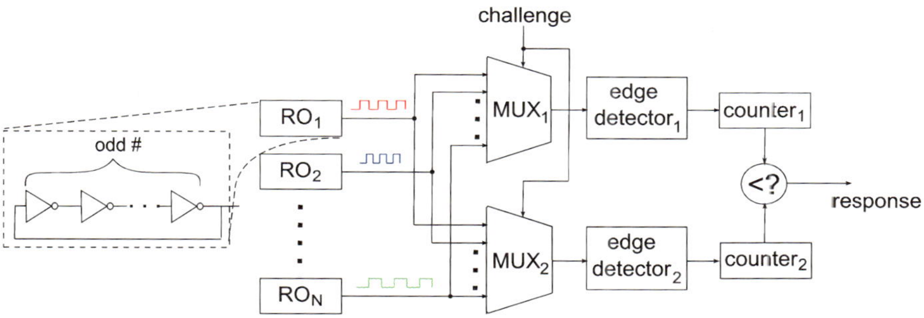

Fig. 12.2 Ring Oscillator (RO) and Ring Oscillator PUF (RO-PUF)

12.2.2.2 Ring Oscillator PUF (RO-PUF)

The Ring Oscillator PUF (RO-PUF) is a delay-based PUF structure that is easier to implement than the Arbiter PUF.

A ring oscillator (RO) circuit consists of an odd number of inverters as shown in Fig. 12.2. The oscillation frequency of an RO is determined by the total delay of its inverters. Due to process variations, the precise frequency is random and IC dependent. An RO-PUF generates signature bits by comparing oscillation frequencies of two or more ROs. A common RO-PUF architecture is shown in Fig. 12.2 [8] and functions as follows. The RO-PUF contains a fixed number of ROs, which are each expected to have slightly different delay/frequency due to process variation. A challenge (input) to the RO-PUF selects two of the ROs. The frequencies of the selected ROs are compared and the response is one bit: a logic 0 (logic 1) if the upper (lower) RO has higher frequency than the lower (upper) RO.

The frequencies of the selected ROs can be obtained quite easily using standard digital components. An edge detector detects the rising edges in output oscillations and a digital counter counts the number of edges over a period of time. A comparator can be used to compare the total number of edges ($\propto$ frequencies) of the two ROs.

12.2.2.3 Aging-Resistant Ring Oscillator (ARO) PUF

The aging-resistant PUF was proposed in [11] to improve reliability which is a major issue in conventional RO-PUF. Like the RO-PUF, a pair of aging-resistant ROs (AROs) are selected by an applied challenge and compared to generate the unique ID. Although the architecture of the ARO-PUF is similar to the RO-PUF, the structures of ring oscillators used in it are different. Figure 12.3a shows the structure of the ARO.

The degradation of an IC due to aging can be attributed to negative bias temperature instability (NBTI) [13, 14] and hot carrier injection (HCI) [15, 16] which is prominent in PMOS and NMOS devices respectively (as described in

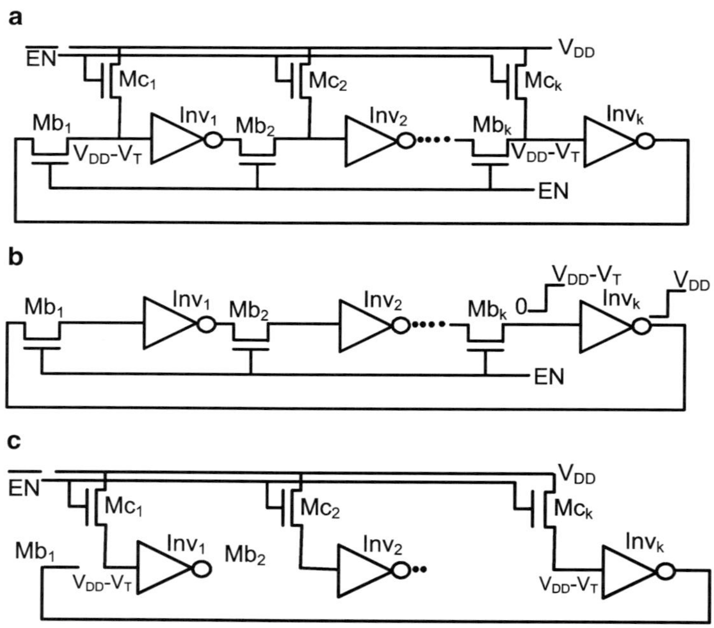

Fig. 12.3 Ring oscillators used in ARO-PUF and operating modes. (**a**) Aging-resistant RO (ARO). (**b**) Oscillation mode. (**c**) Non-oscillation mode

Chap. 3). This degradation depends on threshold voltage, the input stress (DC or AC), size, load, operating temperature, and supply voltage. When a PMOS transistor receives '0' at the gate, it is NBTI stressed and degrades. On the other-hand, when it faces '1', it recovers parts of the NBTI-induced degradation. HCI effect, however, is due to switching between '0' and '1' on an NMOS transistor. The AROs remain in the oscillation mode during the generation of PUF responses (see Fig. 12.3b) and degrades during the measurement. As this time is practically small, the aging experienced by the AROs due to NBTI and HCI are extremely small. Rest of the time the AROs remain in the non-oscillation mode (see Fig. 12.3c) where the gate of all the PMOS transistors are experiencing a logic '1' and there is no switching (no HCI) in the inverter chain.

12.2.2.4 SRAM PUF

An SRAM cell is a circuit that stores one bit of information. A typical SRAM cell consists of cross-coupled inverters (M1,M2 and M3,M4) and access transistors (M5 and M6) as shown in Fig. 12.4. During typical operation, the inverters drive the output nodes (labeled A and A' in Fig. 12.4) to opposing logic values. The SRAM cell stores a LO when $A, A' = 0, 1$V and a HI when $A, A' = 1, 0$V. The access transistors are used to either overwrite or read the bit contained in the SRAM cell.

An SRAM cell exhibits random behavior when reset: (i) when the cell's power supply is off ($V_{dd} = V_{gnd}$), it enters into an unstable state where $A = A' = 0$V; (ii) when power is re-applied to the cell, it transitions from the unstable state into one of the two stable states (LO or HI). The transition to a stable state depends on the parameters (channel length, channel width, threshold voltage, etc.) of each transistor in the cell. Due to manufacturing variations, all these parameters are random and result in a tendency towards one of the stable states after power is reset. An SRAM PUF exploits the random settling behavior of a group of SRAM cells. The challenge (input) to the PUF selects a subset of the SRAM cells to power off. Response bits are the resulting logic values of the selected cells when power is re-applied.

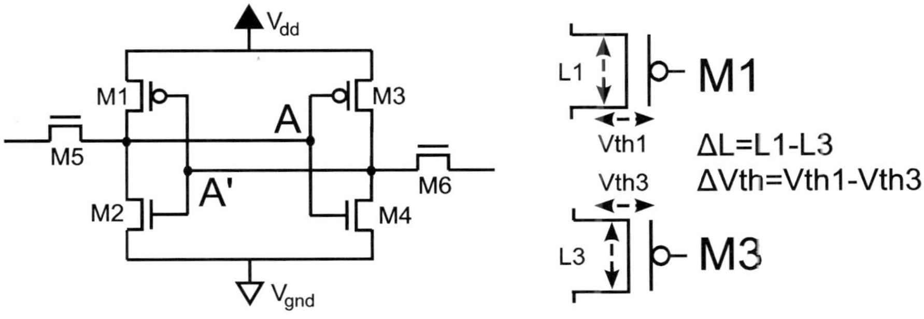

Fig. 12.4 SRAM cell and parameter mismatch between M1 and M3 (ΔL, ΔV_{th})

12.2.3 PUF Quality and Metrics

For use in many practical applications, there are three properties that are very important for PUFs [17]:

i. *Uniqueness:* In order for a PUF signature to be used as a form of identity, any particular challenge should result in a large difference in responses of any two PUF instances (in separate devices). A typical measure for uniqueness is mean *inter-distance* [18]

$$d_{inter}(C) = \frac{2}{k(k-1)} \sum_{i=1}^{k-1} \sum_{j=i+1}^{k} \frac{\text{HD}(r_i, r_j)}{m} \times 100\% \tag{12.1}$$

where $\text{HD}(r_i, r_j)$ is the hamming distance between any two responses r_i and r_j from *different PUFs to the same challenge* C; k is the number of chips/devices in the population under test; and m is the number of bits per response. The optimal $d_{inter}(C)$ is 50 %.

ii. *Reliability:* The response of a particular PUF instance for the same challenge may vary due to temporal variations. However, one desires relatively stable responses so that the PUF can re-generate its key/identifier. A common measure for reliability is mean *intra-distance*. This is calculated by collecting s samples of a response at different operating conditions (supply voltage, temperature, etc.) and computing [18]

$$d_{intra}(C) = \frac{1}{s} \sum_{j=1}^{s} \frac{\text{HD}(r_i, r'_{i,j})}{m} \times 100\% \tag{12.2}$$

where r_i is the nominal response of a challenge C to a PUF; $r'_{i,j}$ is the jth sample of r_i for that *same challenge and same PUF instance*; and m is the number of bits per response. Ideally, $d_{intra}(C) = 0$ which corresponds to no changes in response for challenge C (i.e. perfect reliability).

iii. *Unpredictability:* Since PUFs can be used to generate IDs, PUF responses should be unpredictable/random in order to ensure that the secret remains safe from machine-learning attacks. Several measures of unpredictability have been utilized in the literature. One ad-hoc approach is to determine how well machine learning attacks can be used to model PUF CRPs [17]. More formal metrics such as min-entropy [19] and bit-aliasing [18] measure randomness in the signatures.

12.2.4 PUF Applications in Hardware Security

Silicon PUFs and their associated signatures are convenient for IC identification and authentication. After manufacturing an IC, the vendor can record the challenge-response pairs (CRPs) of its PUF in an enrollment phase. After deployment, a device's identity can be verified at any time by the vendor by applying any challenge from the enrollment phase to the PUF. Since each PUF provides a unique response and the response can only be measured if one has the physical device, the identity of the device is verified as authentic when the response returned is the same as the response recorded during the enrollment phase. To avoid replay (eavesdropping) attacks, the selected challenge should only be used once to identify the device [10]. PUFs are also used in active metering schemes (see Chap. 11) to combat theft, cloning, and overproduction of ICs.

12.2.5 Challenges and Limitations

i. *Reliability*: Reliability is a major concern for most implementations of PUFs today. The response of a PUF must be constant for a given challenge over a wide range of environmental variations, ambient noise, and aging. For RO-PUF, an applied challenge selects a pair from a group of ROs, and the frequency of that pair is compared to generate a one-bit response. Due to the environmental variations and aging impacts, the extent of degradation is different in different ROs.

   Figure 12.5 illustrates how the reliability of a RO-PUF is affected by environmental variations and aging. In a RO-PUF, one can compare the frequency of ROs of a pair selected by given challenge. Let us now consider two ROs, RO1 and RO2 from Fig. 12.2, to generate a single bit of the response. The pair will always generate a reliable bit if the frequency of the RO1 is always higher than the frequency of RO2, regardless of environmental variations and aging.

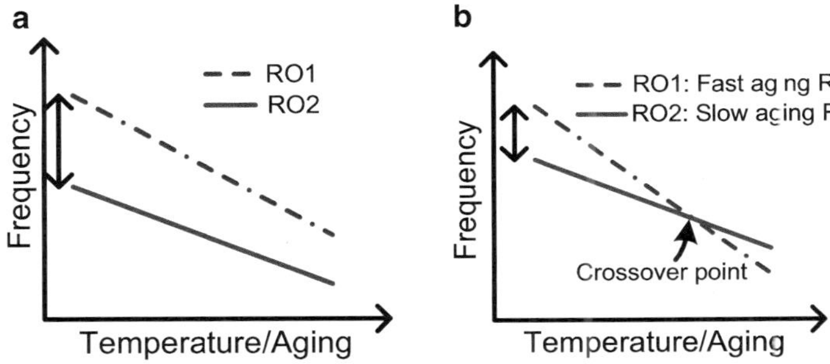

Fig. 12.5 Reliability issue of RO-PUF. (**a**) Stable pair. (**b**) Unstable pair

This pair is known as a stable pair which is shown in Fig. 12.5a. Figure 12.5b shows how a bit could flip beyond a certain operating condition or after a certain amount of aging of the device. At time 0, RO1 is faster than RO2 and the response bit is 1. However, after a device reaches a certain age, RO1 becomes slower than the RO2. This causes the response to be 0 instead of 1 and is called a bit-flip. As temporal variations (changes in voltage supply, changes in temperature, aging) are known to impact the performance, similar reliability issues have been observed in other PUFs such as Arbiter PUF [20] and SRAM PUF [21].

ii. *Component Type*: Today, a large portion of the supply chain is populated by active and obsolete components. The active components are manufactured with previously fixed designs. The obsolete components are no longer manufactured as the OCMs may no longer exist or may have adopted a newer design. There is no opportunity for adding any extra hardware to create a die ID in those designs. In addition, a majority of components in the supply chain belong to small analog and mixed signal categories. In such cases, adding extra hardware for the PUF to the die may not be feasible as it will significantly increase die area.

iii. *Implementation Cost*: This is the area required on the die to implement a counterfeit avoidance measure. For a RO-PUF, to generate a n-bit response we require at least $(n + 1)$ ROs [22]. However, the actual number of ROs are much higher than this [11]. This is also true for the ARO-PUF. In addition, two counters and one comparator are required to implement a RO-PUF. We require n multiplexers and one latch to implement an arbiter PUF to generate n-bit response. The area for either RO-PUF or arbiter PUF is significant for small digital ICs. SRAM-PUF is limited to only those ICs with SRAM. In addition, error correction circuitry (ECC decoder and encoder) is also required for all the PUFs to produce error-free PUF-response.

iv. *Maintenance Cost*: The cost of implementing a PUF would entail storing and maintaining the challenge-response pairs in a secure database, in addition to its area overhead as described earlier. The RO-PUF and ARO-PUF require at least $n * log_2(n + 1)$ bits of challenge to produce n-bit response. Here, $log_2(n + 1)$ represents the selection bits for the multiplexers. The arbiter-PUF requires $n * k$ bits of challenge to produce n-bit response, where k is the number of switch blocks. In addition, we need to store multiple challenges in the database to avoid replay (eavesdropping) attacks and one challenge should only be used once to identify the device [10]. Let us now start with an example where we generate a 128-bit ID and store 100 such IDs in a database for an IC. The number of ROs needed is 129 for RO-PUF. We assume that there are 64 switch blocks in the arbiter PUF. The space required to store such data is $100*128*log(129)\ bits =$ $89.74\ kBits$ for RO-PUFs and $100*128*64\ bits = 819.2\ kBits$ respectively. This is fairly large when we want to store millions of IDs.

12.3 Package ID

The above challenges and limitations lead us to search for alternate methods of creating an ID. Package IDs do not require any modifications in the design or its fabrication process. Hence, package ID may be an ideal solution since it is applicable to active, obsolete, small, and mixed signal components. In the following, we will briefly describe all possible technologies to create a unique package ID.

12.3.1 Encrypted QR Codes

The quick response (QR) code [23, 24], a 2D matrix barcode, is an optical label widely used for product tracking, product identification, and many other document management purposes. The advantage of a QR code is that it can be scanned regardless of scanning direction using a simple hand held device, such as a smartphone [25].

Figure 12.6a shows the structure of a QR code. The QR code is an image consisting of black squares known as modules placed on a white background where each module represents some information about the input text. As the number of

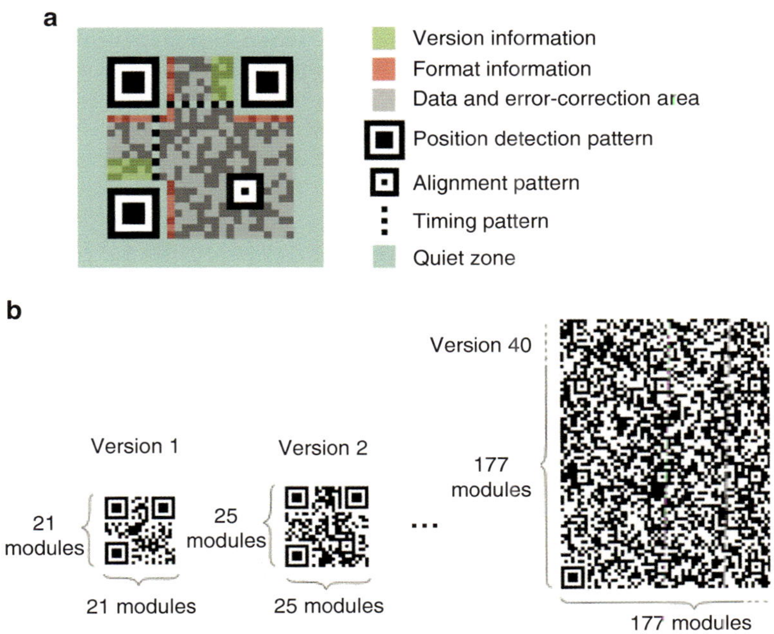

Fig. 12.6 QR code [26]. (**a**) QR code symbol. (**b**) QR Code versions

characters stored in the QR code increases, the size of the modules decreases (see Fig. 12.6b) and it becomes difficult for a smartphone to read the QR code properly due to the limited resolution of the camera. There are 40 different versions of QR codes available today, depending on the size of the symbol.

The authors in [27] proposed an authentication method using photon-counting encryption implemented with phase encoded QR codes. The marking on the package of an IC is converted to an optically encoded QR code. The encryption of the QR code image is performed using the full phase double-random-phase encryption with photon-counting [28] to prevent the copying of the information by the counterfeiter. An iterative compression technique based on Huffman coding [29] is then used to compress the photon counting encrypted image to reduce the size of the QR code. Note that the lower version of QR code can easily be scanned by any commercially available smartphones.

The authentication can be easily performed by scanning the QR code with a commercially available smartphone. The encrypted, scanned data is decompressed and then decrypted. Image recognition algorithms such as nonlinear correlation filters can be used to verify the decrypted image against the primary image for authentication.

12.3.2 DNA Markings

Deoxyribonucleic acid (DNA) provides a unique identification where botanical DNA taggants are applied on the package. This essentially tags the electronic components to trace them throughout the component supply chain. Plant DNA is scrambled to create new and unique genetic sequences, and these sequences of DNA are integrated with inks. These inks are then applied on the packages of the IC at the end of the packaging process. Figure 12.7 shows the DNA extraction and application process by the Signature® DNA program of Applied DNA Sciences (ADNAS) [30]. Recently, the DOD mandated [31] that DNA marking be placed on the components in order to track them throughout the supply chain.

DNA marking provides unique protection against copying of the ID created by the DNA taggants. Unlike other marking techniques, the marks created by the DNA cannot be simulated, copied and/or reproduced. Any attempts to remove the DNA

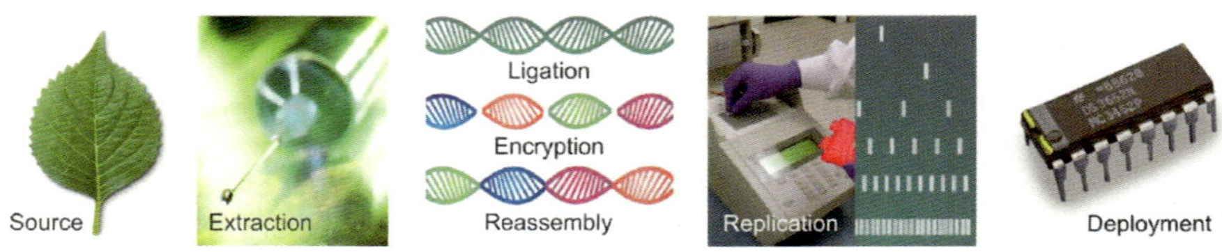

Fig. 12.7 The creation and deployment of DNA taggants [30]

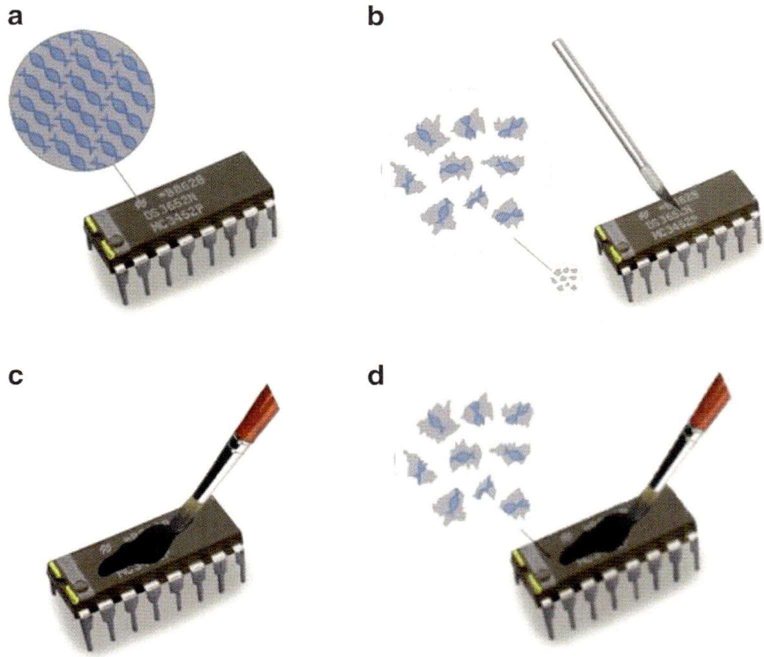

Fig. 12.8 Safeguard against recycling process [32]. (**a**) Original DNA mark. (**b**) DNA marks are removed by scrapping. (**c**) Blacktopping covers DNA marks with new material. (**d**) Reused DNA marks

from the package by the recycling process, will damage the DNA mark. When the DNA marks are applied to the package, it creates an ordered structure. During the recycling process which involves sandblasting and blacktopping/resurfacing, the ordered DNA pair is either damaged or covered by new material. The counterfeiter would not be able to reuse the DNA collected from the component since it is already damaged. Figure 12.8b–d show all these scenarios.

The counterfeit prevention authentication (CPA) program by ADNAS, is designed to track authentic components throughout the supply chain. Red authentic mark is used by the OCMs while green and yellow marks are used by the legally authorized distributors and distributors. Figure 12.9 shows the CPA program. Once the components are received, the authentication includes checking whether the ink fluoresces under UV light, and then, sending a sample of the swab ink to a ADNAS forensic laboratory to verify that the DNA is in the database of valid sequences [33].

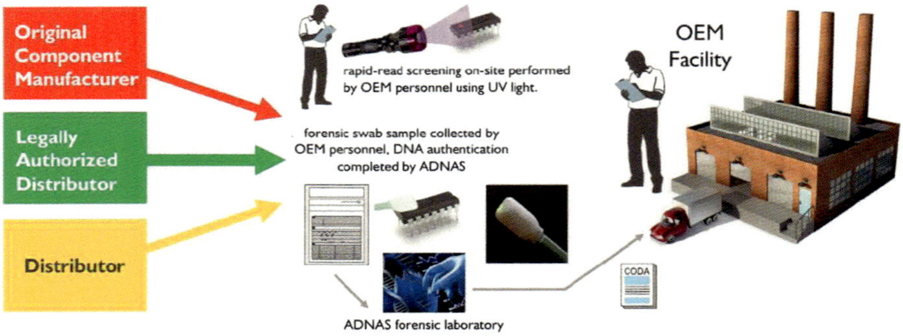

Fig. 12.9 The counterfeit prevention program by ADNAS [33]

12.3.3 Nanorods

IBM researchers introduced gold nanorods on a surface using a simple printing process [34]. In this technique, a microscopic pattern is created by growing an array of nanospheres into nanorods that are less than 100nm long. Each time the process is repeated, the same pattern is created, but the exact angle and length of each individual nanorod varies, so that each set of nanorods is distinct. After the array of nanorods is grown, it is applied to a chip using a specialized printer. A chip with gold nanorods on its surface can be authenticated by comparing the overall pattern and visual properties of each nanorod to a database.

Along with nanorods, IBM researchers also created different patterns using red, green and blue fluorescent spheres [35]. Figure 12.10 illustrates a fluorescence microscope image (channel overlay), which consists of 1-μm diameter fluorescent polystyrene spheres assembled in a corner array. Here, the color of the sphere is not predictable even though the position of single particles is known. It is impossible to reproduce the same colored arrays as the number of possible color combinations is considerably large.

12.3.4 Capacitive (Coating) Physical Unclonable Functions

Capacitive (coating) PUF, introduced in [36], can potentially be a suitable candidate for creating a package ID. In coating PUF, the IDs can be generated from the capacitance measurements of an array of metal sensors directly beneath the passivation layer and a coating on top of the passivation layer containing many randomly distributed particles with different dielectric constants.

Figure 12.11 shows the structure of a coating PUF as proposed in [36]. A network of metal wires is laid out in a comb shape just beneath the passivation layer of an IC. The space between and above the comb structure is filled with a coating consisting

Fig. 12.10 Fluorescence microscope image of 1-μm diameter fluorescent polystyrene spheres [35]

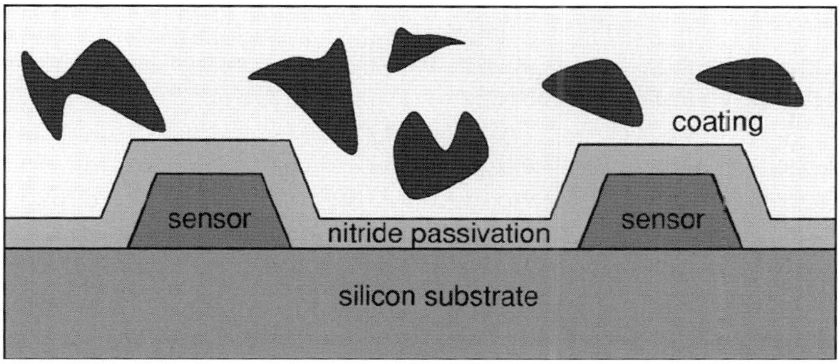

Fig. 12.11 Structure of a coating PUF [36]

of aluminophosphate. The coating is doped with random dielectric particles with different sizes and shapes and have a different dielectric constant than the coating matrix. Note that in contrast to the PUFs discussed in Sect. 12.2.1, the coating PUF is not implemented in the silicon die. Hence, we do not consider it as a die ID.

As discussed in Sect. 12.2.1, there are two steps for using PUFs for IC authentication: enrollment and verification. In enrollment, a finite number of input combinations (challenges) are chosen randomly and the corresponding output (responses) of the PUF are collected and stored in a database. During verification, one or more previously enrolled challenges are applied to the PUF and the responses corresponding to those challenges are compared with the stored responses. If the responses match those corresponding to the device under authentication within a certain threshold, then the device is considered authentic.

In this coating PUF, a challenge corresponds to a voltage of a certain frequency and amplitude applied to the two terminals of the matrix. The capacitance at various location varies due to the random density of dielectric particles. The capacitance values at different locations are converted into a bit string that can be used as a unique die ID. It is reported that this structure can reliably produce an ID with length in the order of 100 bits/mm^2.

12.3.5 Challenges and Limitations

There are several challenges associated with the existing package ID generation techniques. Table 12.1 presents a comparison of all the different technologies in terms of the requirements of package IDs discussed in Sect. 12.1. We have assigned a score of high, medium, or low depending on effectiveness.

Reliability is a major issue that must be overcome for many of these techniques. The reliability of QR codes on the package of a component has not been verified nor has the technique been accepted by the semiconductor industry. This is also true for nanorods and coating PUFs. The authors claim that the DNA marks are stable up to 250°C, stable under UV ray, X-Ray, and Y-Ray, and also pass the solvent test [32].

Due to the optical encryption used while applying QR code [27] on the package of a component, the package ID represented should be unique and unclonable. The uniqueness for the coating PUF and nanorods should be high as well. For DNA markings, a cursory authentication includes checking whether the DNA mark is present under the UV light. This could easily be circumvented by the counterfeiters because one only needs to mimic the material of the marker to produce the same colored light [37]. The detailed validation of the DNA requires a forensic lab and is fairly time consuming. Thus, it can only be performed on a sampling basis.

Manufacturability is another major problem that needs to be overcome in the near future. Today, none of the above four technologies can be implemented at the high volume required to support the entire semiconductor industry. The cost of producing markings is not verified for most of these technologies. According to SIA [37], the implementation for the DNA markers will greatly increase the overall manufacturing costs as it requires the modification of long-standing stable manufacturing flows. The same report also mentioned that DNA marking has not been subjected to the standard reliability qualification and failure prevention tests that are currently in practice for the semiconductor industry. From a usability perspective, QR codes provide better and easier authentication through the use of widely available smartphones.

Table 12.1 Assessment of different IDs

IDs	Reliability	Uniqueness	Unclonable	Manufacturability	Cost effectiveness	Ease-of-Use
QR codes	Not verified	Medium	Medium	Not verified	Not verified	High
DNA markings	Low	Low	Low	Low	Low	Medium
Nanorods	Not verified	High	High	Not verified	Not verified	Medium
Coating PUF	Not verified	High	High	Not verified	Not verified	Medium

12.4 Limitations of Chip IDs for Different Counterfeit Types

Package and die IDs suffer from several limitations and challenges. Most notably, the unclonable nature of IDs provide protection against some counterfeit types, but not all of them. In the following, we will describe the detection and avoidance of different counterfeit types by using die and package IDs and highlight the possible challenges with using IDs for each counterfeit type,

i. Recycled: Die IDs cannot be used to detect recycled ICs, but package IDs are useful. The recycled ICs will be detected when the counterfeiter removes the old marking on the package by sandblasting or other process and then resurfaces and remarks the package with new markings. As the markings are unclonable, the counterfeiter cannot reproduce the same ID. However, recycled ICs cannot be detected if the counterfeiters skip the steps for the removal of old markings.
ii. Remarked: As long as grade, manufacturer, etc. are linked to certain IDs, many of the unclonable IDs could detect remarking. In the case of package IDs, they are also impossible to recover once the marking is removed. Hence, remarked ICs can definitely be detected through package IDs.
iii. Overproduced: The detection of overproduced ICs are possible as the IDs for the overproduced ICs are not registered in OCM's database. Essentially, the untrusted foundries can still sell these ICs without the knowledge of the design houses.
iv. Defective/Out-of-Spec: These ICs cannot be authenticated by checking the IDs as they may all hold valid IDs.
v. Cloned: The unclonable nature of IDs provide protection against cloning. However, similar to overproduction, ICs fabricated with a cloned design can still be sold without the knowledge of the design house.
vi. Forged Documentation: These ICs can easily be detected as there will be a mismatch of information between the documents received and the IDs.
vii. Tampered: IDs cannot detect tampered ICs when tampering is performed at the die level, i.e., before packaging of ICs as all these tampered ICs could have valid IDs.

In summary, we conclude that remarked, and forged documentation counterfeit ICs will definitely be detected by implementing unclonable IDs. Some portions of recycled, cloned, and overproduced ICs will also be detected. However, it is certain that one cannot detect defective/out-of-spec and tampered ICs by using chip IDs.

12.5 Summary

In this chapter, we presented chip IDs as a technique to track and trace components as they move through the component supply chain. Traceability has become particularly important as ICs are now manufactured and assembled all across the world, leading to concerns regarding the authenticity of the components as well as the manufacturers themselves.

Die IDs are popular for large digital ICs and are resistant to various attacks when PUFs are used to produce the IDs. We have introduced four different types of PUFs to generate unclonable IDs. Arbiter PUF was the first PUF proposed to generate an unclonable ID. However, as we mentioned earlier, there are several limitations to designing a robust arbiter PUF. To generate a correct ID, it is necessary to satisfy the setup time and hold time requirements of the D-latch, and both paths for each stage in the arbiter PUF must be perfectly symmetric. It has also been shown that after observing a number of CRPs, simple machine-learning techniques can be used to predict PUF responses, which allow attackers to determine an ID without being in possession of an IC. The RO-PUF generates IDs while solving the limitations of the arbiter PUF. Reliability is a major challenge for RO-PUFs as the bits in the ID change over time due to aging. ARO-PUF was proposed to design a reliable RO-PUF and solve this issue. However, the area overhead for both RO-PUF and ARO-PUF still remains a major issue that must be addressed in the near future. SRAM-PUFs generate IDs based on the random settling behavior of a group of SRAM cells when a challenge selects those cells. In the SRAM PUF as well as all the other PUF architectures that have been suggested, generation of a stable PUF response still remains a major challenge. Reliability issues aside, existing PUFs are limited to large digital ICs and thus inapplicable to analog, small discrete, etc. components in the supply chain. In addition, all the PUFs require a large secure storage to record the challenges to produce unique IDs during IC authentication. Maintaining such large databases and giving public access through a network still remain as one of the major bottlenecks in PUF implementation.

Package IDs can be marked onto any type of components, regardless of whether they are new, active or obsolete, small, medium or large, or analog, digital or mixed-signal. They contain information such as date and lot code, manufacturer ID, country of origin and ID numbers to identify each component uniquely. We also pointed out the key features required in package IDs, such as uniqueness, unclonability, reliability, cost-effectiveness, and others. Four unique techniques for producing package IDs were introduced. Encrypted QR codes, which can be implemented by using 2D matrix barcodes, seem to be a promising candidate for creating a unique package ID, but require some additional testing with regards to reliability, manufacturability, and cost, and are yet to be implemented widely by the industry. DNA sequences extracted from plants can also be used to stamp a unique ID on IC packages, which become distorted on any attempt to tamper (possibly during recycling and remarking) and render the DNA-marked IC as suspect. Nanorods were also introduced where microscopic patterns, each with unique size and orientations of nanorods, are printed onto the packaging or die of an IC. Finally, coating PUFs were suggested for producing package ID on a die by utilizing the unique capacitance measurements at various locations in an IC coated with a doped aluminophosphate layer.

Finally, die and package IDs alone cannot provide enough protection against all counterfeit types. Defective/out-of-spec counterfeit components cannot be detected by chip IDs, as these components could possess a valid ID. Further, if recycled ICs preserve their chip IDs, they cannot be identified as counterfeit if they re-enter the supply chain.

References

1. SAE, Counterfeit electronic parts; avoidance, detection, mitigation, and disposition, 2009, http://standards.sae.org/as5553/
2. U. Guin, D. DiMase, M. Tehranipoor, Counterfeit integrated circuits: Detection, avoidance, and the challenges ahead.," J. Electron. Test. **30**(1), 9–23 (2014)
3. U. Guin, D. Forte, M. Tehranipoor, Anti-counterfeit techniques: from design to resign, in *Microprocessor Test and Verification (MTV)*, 2013
4. Departnent of Defense, Performance Specification: Hybrid Microcircuits, General Specification For, 2009, http://www.dscc.dla.mil/Downloads/MilSpec/Docs/MIL-PRF-38534/prf38534.pdf
5. K. Arndt, C. Narayan, A. Brintzinger, W. Guthrie, D. Lachtrupp, J. Mauger, D. Glimmer, S. Lawn, B. Dinkel, A. Mitwalsky, Reliability of laser activated metal fuses in drams, in *Proc. of IEEE on Electronics Manufacturing Technology Symposium*, 1999, pp. 389–394
6. N. Robson, J. Safran, C. Kothandaraman, A. Cestero, X. Chen, R. Rajeevakumar, A. Leslie, D. Moy, T. Kirihata, S. Iyer, Electrically programmable fuse (efuse): From memory redundancy to autonomic chips, in *CICC*, 2007, pp. 799–804
7. B. Gassend, D. Clarke, M. Van Dijk, S. Devadas, Silicon physical random functions, in *Proc. CCS* (ACM, 2002), pp. 148–160
8. G. Suh, S. Devadas, Physical Unclonable Functions for device authentication and secret key generation, in *Proc. DAC*, 2007, pp. 9–14
9. J. Guajardo, S. Kumar, G. Schrijen, P. Tuyls, FPGA intrinsic PUFs and their use for IP protection, in *Proc. CHES*, 2007, pp. 63–80
10. R. Maes, I. Verbauwhede, Physically Unclonable Functions: A study on the state of the art and future research directions. Towards Hardware Intrinsic Secur., pp. 3–37, 2010
11. M. Rahman, D. Forte, J. Fahrny, M. Tehranipoor, Aro-puf: An aging-resistant ring oscillator puf design, in *Design, Automation and Test in Europe Conference and Exhibition (DATE), 2014*, March 2014, pp. 1–6
12. S. Morozov, A. Maiti, P. Schaumont, An analysis of delay based puf implementations on fpga. Reconfig. Comput. Architect. Tools Appl. 382–387 (2010)
13. M. Alam, S. Mahapatra, A comprehensive model of pmos nbti degradation. Microelectron. Reliab. **45**(1), 71–81 (2005)
14. S. Bhardwaj, W. Wang, R. Vattikonda, Y. Cao, S. Vrudhula, Predictive modeling of the nbti effect for reliable design, in *Proc. of IEEE on Custom Integrated Circuits Conference*, September 2006, pp. 189–192
15. K.-L. Chen, S. Saller, I. Groves, D. Scott, Reliability effects on mos transistors due to hot-carrier injection. IEEE Trans. Electron Dev. **32**(2), 386–393 (1985)
16. S. Mahapatra, D. Saha, D. Varghese, P. Kumar, On the generation and recovery of interface traps in mosfets subjected to nbti, fn, and hci stress. IEEE Trans. Electron Dev. **53**(7), 1583–1592 (2006)

17. I. Verbauwhede, R. Maes, Physically unclonable functions: manufacturing variability as an unclonable device identifier, in *Proc. GLSVLSI* (ACM, 2011), pp. 455–460
18. A. Maiti, P. Schaumont, Improved ring oscillator puf: An fpga-friendly secure primitive. J. Cryptology, 1–23 (2011)
19. S. Katzenbeisser, Ü. Kocabaş, V. Rožić, A. Sadeghi, I. Verbauwhede, C. Wachsmann, PUFs: Myth, fact or busted? A security evaluation of Physically Unclonable Functions (PUFs) cast in silicon. Proc. CHES, 283–301 (2012)
20. D. Lim, J. Lee, B. Gassend, G. Suh, M. van Dijk, S. Devadas, Extracting secret keys from integrated circuits. IEEE Trans. Very Large Scale Integration (VLSI) Syst. **13**(10), 1200–1205 (2005)
21. K. Xiao, M. Rahman, D. Forte, Y. Huang, M. Su, M. Tehranipoor, Bit selection algorithm suitable for high-volume production of sram-puf, in *2014 IEEE International Symposium on Hardware-Oriented Security and Trust (HOST)*, May 2014, pp. 101–106
22. A. Maiti, P. Schaumont, Improved ring oscillator puf: An fpga-friendly secure primitive. J. Cryptology **24**(2), 375–397 (2011)
23. D. Wave, Answer to your questions about the QR Code, http://www.qrcode.com/en/
24. ISO/IEC 18004:2006, Information technology - Automatic identification and data capture techniques - QR Code 2005 bar code symbology specification, (2006) http://www.iso.org/iso/catalogue_detail?csnumber=43655
25. E. Ohbuchi, H. Hanaizumi, L. Hock, Barcode readers using the camera device in mobile phones, in *2004 International Conference on Cyberworlds*, Nov 2004, pp. 260–265
26. D. Wave, QR Code Essentials. [Online]. Available: http://www.nacs.org/LinkClick.aspx?fileticket=D1FpVAvvJuo%3D&tabid=1426&mid=4802
27. A. Markman, B. Javidi, M. Tehranipoor, Photon-counting security tagging and verification using optically encoded qr codes. Photonics J. IEEE **6**(1), 1–9 (2014)
28. E. Pérez-Cabré, H.C. Abril, M.S. Millán, B. Javidi, Photon-counting double-random-phase encoding for secure image verification and retrieval. J. Optics **14**(9), 094001 (2012). [Online]. Available: http://stacks.iop.org/2040-8986/14/i=9/a=094001
29. D. Huffman, A method for the construction of minimum-redundancy codes. Proc. IRE **40**(9), 1098–1101 (1952)
30. Signature DNA, http://www.adnas.com/products/signaturedna
31. U.S. Defense Logistics Agency, Dna authentication marking on items in fsc 5962, August 2012. [Online]. Available: https://www.dibbs.bsm.dla.mil/notices/msgdspl.aspx?msgid=685
32. J.A. Hayward, J. Meraglia, DNA Marking and Authentication: A unique, secure anti-counterfeiting program for the electronics industry, Oct. 2011
33. Applied DNA CPA Program, http://www.adnas.com/CPA
34. C. Kuemin, L. Nowack, L. Bozano, N.D. Spencer, H. Wolf, Oriented assembly of gold nanorods on the single-particle level. Adv. Funct. Mater. **22**(4), 702–708 (2012)
35. IBM Research, Nanorods take down counterfeiters: IBM scientists create nano-sized patterns to thwart forgeries, http://www.research.ibm.com/articles/nano-counterfeit.shtml
36. B. Skoric, S. Maubach, T. Kevenaar, P. Tuyls, Information-theoretic analysis of capacitive physical unclonable functions. J. Appl. Phys. **100**(2), 024902 (2006)
37. Semiconductor Industry Association (SIA), *Public Comments - DNA Authentication Marking on Items in FSC5962*, Nov 2012

Index

M. Tehranipoor et al., *Counterfeit Integrated Circuits*,
DOI 10.1007/978-3-319-11824-6